T0094128

HOW TO WIN FRIENDS and INFLUENCE FUNGI

HOW TO WIN FRIENDS and INFLUENCE FUNGI

Collected Quirks of Science, Tech, Engineering, and Math from Nerd Nite

Dr. Chris Balakrishnan
and Matt Wasowski

Illustrations by Kristen Orr

ST. MARTIN'S PRESS NEW YORK

First published in the United States by St. Martin's Press, an imprint of St. Martin's Publishing Group

 Printed in the United States of America. For information, address St. Martin's Publishing Group, 120 Broadway, New York, NY 10271.

www.stmartins.com

Designed by Jonathan Bennett

The Library of Congress Cataloging-in-Publication Data is available upon request.

ISBN 978-1-250-28834-9 (hardcover)

ISBN 978-1-250-28835-6 (ebook)

First Edition: 2024

10 9 8 7 6 5 4 3 2 1

To our parents for their inexhaustible support

To Jessica and Sarah for . . . everything

To Jane, it is really fine if you opt not to become a bird scientist

xiii *Introduction*

1 **Creature Features**

3 Camel Spiders: The Rumors of My Size Have Been Greatly Exaggerated
by Forest Ray, PhD

7 Military Marine Mammals: Dolphins So Smart They Should Give Their Own Nerd Nite Presentation
by Laura Chaibongsai

9 Sex Catapults!
by Ben Taylor

13 Cephalopods: The Impossibly Awesome Invertebrates
by Aerie Shore

17 Stomatopods: Why Is My Thumb Bleeding and My Mind Blown?
by Peter C. Thompson, PhD

21 Finding Nemo('s Sex): Sex Change and Gender Roles in Anemonefishes
by Jann Zwahlen

25 **Mmm . . . Brains**

27 It's Not You, It's Misophonia
by Dr. Jane Gregory

31 Sex, Drugs, and Happiness
by Melissa Blundell-Osorio

36 Don't Trust Your Brain: Why Foreign Accents Are All in Your Head
by Mari Sakai, PhD

40 Lessons from *The Oregon Trail*
by Elizabeth Russell, PhD

44 Synesthesia: Hearing Colors and Tasting Sounds
by Rebecca Woods, PhD

49 Brain on a Chip: The Ethics of Brain Experimentation
by Max Jackson

52 How We Become Disgusting (Some More Than Others)
by Dr. Richard Firth-Godbehere

57 **Bodily Fluids**

59 To Boldly Go: Dealing with Poop and Pee in Space
by Brendan Byrne

62 Milk! You'll See It Everywhere
Once You Know What to Look For
by Kristen Orr

65 Triclosan: It's Not the Bacteria but the Soap
That's Going to Kill You!
by Dr. Graeme R. A. Wyllie

69 Lost: Bladder Control. Reward for Safe Return.
by Maria Jantz

72 Microbes Can Save You, Kill You, or Just Give You the Poops
by Ariane L. Peralta, PhD

75 Runoff: What Is It and Why Even Notice It?
by Evelyn M. Zornoza, RLA

81 **Doing It**

83 Hot or Not? How to Be a Perfect 10; or, How to Manipulate
Perceptions of Physical Attractiveness
by Lillian Park, PhD

86 Dating Tips from the Animal Kingdom:
What to Wear and How to Flaunt It
by Kaci Fankhauser

90 Dating as a Data Nerd
by Tristan Attwood

95 10 Things You Didn't Know About Sex . . . Education
by Anna V. Eskamani

99 Going Ape for Pansexual Primates
by Natalia Reagan

103 Smells and the Microbiome: Are Microbes Controlling Your Sex Life?
by Dr. Jenny Bratburd

107 **Health and (Un)Wellness**

109 Maggot Therapy; or, How I Learned to Stop Worrying and Love the Bugs
by Avir Mitra, MD

113 Happy 13th Birthday, Nerd Nite! Now Get Yourself to an Adolescent Medicine Doctor
by Dr. Nancy Dodson

117 What Your DNA Says About You
by Shweta Ramdas

121 The Science of the Hangover
by Paula Croxson, DPhil

127 You and Your Microbiome: Say Hello to Your Little Friends
by Rebecca B. Blank, MD, PhD

133 Penis or Vagina? 'Tain't That Simple!
by Krista A. McCoy, PhD

136 The Modern Study of Genetics Is Full of Twists and Turns
by Dr. C. Brandon Ogbunu

141 **Pathogens and Parasites**

143 Everything You Always Wanted to Know About Birds
by Dr. Christopher N. Balakrishnan

146 Pigeons, Cannibals, and Vaginas: The Story of My Favorite Parasite
by Andrew Peters

151 What Birds Can Teach Us About the Impending Zombie Apocalypse
by James S. Adelman

155 Zombies Are Real and You Might Be One
by Jeremy N. Kay, PhD

159 Hacking the Antiviral Immune Response
by Dr. Ebony Monson

161 Human Parasites (No, Not Your Mooching Roommate)
by Dr. John Dodson

163 **Death and Taxes (But Really, Just Death)**

165 Monarch the Bear: A Tale of Tycoons, Taxidermy, and the California Flag
by Kelly Jensen

168 How to Not Destroy Ourselves: Lessons from Sci-Fi
by Dr. Ali Mattu

172 Mass Extinction
by Dr. Anirban Bhattacharjee, Thomas A. Shiller II, PhD, and Dr. Sean Graham

176 How and Why Cancer Happens; or, If You Live Long Enough You're Going to Die
by Dr. Kerry P. Donny-Clark

179 Algae Apocalypse: The Most Important Slime
by Lewis Weil and Rose-Anne Meissner, PhD

183 **Space, the Big and the Beautiful**

185 Bullshit in Space: An Astronomical Adventure Through Cosmic Misinformation
by Ralph Crewe

189 Preparation A: Our Relief Against Severe ASSteroids
by Derek Demeter

191 Life Under the Ice of Europa
by Guillermo Garcia Costoya

194 Artificial Gravity in Science Fiction
by Erin Macdonald, PhD

199 Sky Rockets in Flight, Asteroids Delight: Asteroid Mining for Science, Profit, and Fun!
by Dr. Zoe Landsman

201 Better than NASA: Canada's Sample of an Asteroid; or, the Untold Story of the Tagish Lake Meteorite
by Christopher D. K. Herd, PhD

205 The Telescope That Blew Everyone's Mind . . . Part Two!
by Joel D. Green, PhD

209 **Tech (High and Low)**

211 "They're Putting Acid in Our Food!": The Everyman's Guide to Thwarting Fear and Understanding GMOs
by Tracy Kurtz

214 What I Learned About Dating Apps (Generally) After I Spent Five F**king Years Studying Them for a PhD
by Dr. Nicolette Wei Mei Wong

217 Adventures in Human-Powered Flight
by David Donaldson, MCATD, CTDP, CMP, PMP, PRP

221 What Does Google See?
by Michelle Henderson

226 Becoming a Cyborg Through Disability: Building Prosthetic Limbs
by Rafaela Libano

230 Machine Learning for a Free and Open Internet
by Dr. Brandon Wiley

234 How to Win Friends and Influence Bacteria
by Sarah Richardson, PhD

239 Why Nuclear Fusion Would Be Awesome—If We Get It to Work
by Dr. Matt Moynihan

243 **Math Is Fun**

245 A Tea Test Tempest
by Sam Kean

248 The Mathematics of Gossip
by Izabel Aguiar

251 From Bach to Tool: The Secret Math Behind Music Theory
by Alexander Brewer

257 Getting to Know Infinity
by Dr. Zajj Daugherty

261 Math for a Better City
by Eliza Harris Juliano

265 A Little "Bit" of Cryptography
by Marshall Swatt

269 **Careers**

271 Veterinary Confidential
by Jessica Girard, DVM MS

274 Chindogu: The Japanese Art of Unuseless Inventions
by Josh Manning

277 Caskets, Corpses, and Biers, Oh My!: A Brief Look at Death Care and the History and Science of Embalming in the US
by Deanne M. Rugani

282 Wildlife Detectives: The Science and Stories of "Animal CSI" in Investigating and Solving Wildlife Crime
by Dr. Rebecca N. Johnson AM

285 Cut It Off!: A Civil War Amputation
by John Lustrea

289 Fermentation: A Cultural Story
by Amy Oxenham

294 Fire: Of Flames and Friendship
by Lee M. Bishop, PhD

297 ***Our Beloved Nerd Nite Bosses***

303 ***Acknowledgments***

INTRODUCTION

For 21 years, Nerd Nite has delivered to live audiences the most interesting, fun, and informative presentations about science, history, the arts, pop culture—you name it (we're modest, too)—as there hasn't been a rabbit hole that our army of presenters hasn't been afraid to explore. Curious about the science of *The Simpsons* or the cuisine of ancient Egypt? We've had presentations about those. Cyborg feminism? Check. Math feuds? Check. The history of the jumpsuit? Check! Nerd Nite has made hundreds of thousands of people in more than 200 cities across the globe a little bit smarter, and tipsier, by bringing them in-person, fun-yet-informative presentations while they drink along.

And the whole time, we toyed with various ways to expand Nerd Nite. TV shows. Documentaries. Rock operas. Puppets. Interpretive dance. If you can think of it, we batted it around. But we think we've stumbled on the optimal method of spreading Nerd Nite goodness—good ol' print. Timeless, classic, tree-murdering print.

Though we have a YouTube channel that features a few dozen nicely edited Nerd Nite presentations from around the world, we've still only scratched the surface when it comes to sharing our presenters' insanely fun-yet-informative content. For example, if you weren't able to attend the in-person Nerd Nite Miami about *Where Shark Babies Come From*, it doesn't mean you should never be able to learn about its contents ever again. That's just mean.

So here we are. Alone. Just us. Hello.

"But why a book?" you may ask. Or "Why now?" you may wonder. We'll tell you.

First, because you're smart. We know that. You're insatiably curious. We know that, too. You like to laugh and maybe even prefer to drink while you learn. We definitely know that. And most important, we know you're underserved when it comes to heavy, thick books that cater to your delightful wit and natural curiosity about the world. Though it may be easy to find information about animal sex in one place, birdsong in another, the first vaccine in yet another, and the science of the hangover in a fourth, it's much more difficult to find in-depth, quirky content about multiple scientific subjects in one spot. Therefore, we think this book will fill that void of underservedness. With plenty of quirkiness and silliness along the way.

Second, the problem with which we've wrestled since Nerd Nite began expanding, particularly internationally, is how to bring such a diverse and broad array of content to everyone across this big blue marble of ours. We've been told countless times that most Nerd Nite presentations in one city would be perfect for fellow nerds in all our other locales. For instance, in March 2022, Nerd Nite New York featured a presentation called *Maggot Therapy: How I Learned to Love Maggots in the ER*, and in April 2022 Nerd Nite Pittsburgh showcased *What I Learned About Dating Apps (Generally) After I Spent Five F**king Years Studying Them for a PhD*—but we were in Alexandria, Virginia, and Los Angeles, respectively, unable to travel the thousands of miles necessary to witness these presentations in their natural habitats. And we've never even come close to figuring out how we could take a few weeks off to travel to our Nites in Amsterdam, Melbourne, Tokyo, London, and seemingly everywhere in Germany. I'm sure many New Yorkers and Madisonians were equally disappointed that they couldn't visit Nerd Nite San Francisco in May 2022 to learn about *Godzilla: History, Biology, and Behavior of Hyper-Evolved Theropod Kaiju*.

And even though we sporadically videotape (yes, we realize we just wrote "tape"—we're now both comfortably middle-aged Gen Xers) some of our presentations, we still have a difficult time disseminating all the content from the nearly 200 of them that Nerd Nite hosts each month. Therefore, we believe *How to Win Friends and Influence Fungi: Collected Quirks of Science, Tech, Engineering, and Math from Nerd Nite* will allow us to share the exhilarating content from our presenters worldwide.

As an aside, almost since the inception of Nerd Nite in Boston in 2003, we've had two dueling taglines: "Be there and be square" and "It's like the Discovery Channel™ with beer." So now our challenge is to figure out how we can ensure that these identities translate from an in-person event to the printed page. Because you can't actually "be here" (and thereby "be square") when leafing through this book, we must defer to our other tagline. So in the spirit of "It's like the Discovery Channel™ with beer," we certainly wouldn't mind if you help yourself to a cocktail or three while reading these fine pages. In fact, this very book might even make a nice coaster or place mat on the off chance you can actually put it down.

We're also excited that this book will showcase 70 of our favorite nerds from around the world, as Nerd Nite has always prided itself on being a launching pad for emerging folks: About 98 percent of our presenters were either grad students or young professionals in their twenties and thirties when they first presented. We always wanted to be a platform for the next generation of experts, not for household names. Though, interestingly, now that we've been around for two decades, there are innumerable instances of our once-fledgling nerds who have gone on to achieve the tremendous success we always knew they would. We've had past presenters become bestsellers, award-winning documentarians, museum curators, doctors, lawyers, all-star podcasters, politicians, and eminent researchers, but we'll forever remember them as enthusiastic, wise, wily weirdos who simply wanted to passionately present in front of a slightly tipsy audience of like-minded folks. We're even responsible for dozens of marriages and dozens of children—yes, we're grooming the next generation of nerds already! Ah, Nerd Nite, the bespectacled face that launched a thousand ships.

But most of all, we want this book to retain and build upon the spirit of our face-to-face events. We want its stories, lessons, jokes, infographics, and illustrations to be as fun, irreverent, challenging, approachable, and smart as the presentations you've grown to love.

So now, somewhere between *The World Almanac*, *MAD* Magazine, *Science for Dummies*, your university's alumni magazine, and probably not the Bible, we proudly bring you *How to Win Friends and Influence Fungi: Collected Quirks of Science, Tech, Engineering, and Math from Nerd Nite*. Shout-out to Dr. Sarah Richardson, whose contribution in the Tech (High and Low) chapter inspired our title (along with Dale Carnegie, of course). Cheers!

Excitedly, nervously, and still slightly nauseously yours,

Chris and Matt (two easily influenced fun-guys . . . get it?!)

Creature Features

"People love to learn about weird creatures on Earth." This isn't quite one of Newton's laws, but it might as well be. Yes, we realize the exact phrasing still needs work, but it doesn't change the fact that people love animals. It doesn't matter if your interest is in watching them, petting them, feeding them, drawing them, teasing them, training them, talking to them, eating them, hunting them, or preventing people from eating them or hunting them, curiosity about the natural world abounds. This is true even if you'd rather watch David Attenborough or *Wild Kratts* on TV than actually set foot in a jungle. In fact, we're flattered that you're even taking a few minutes from watching cat videos on YouTube right now to read this sentence. Who among us can resist a tale of a tiny shrimp as powerful as a .22-caliber bullet, spider sex catapults, spiders of unusual size, or sex-changing fish? Because of this universal adoration of our planet's weird and wonderful fauna, Nerd Nites throughout the land have always featured a heavy dose of weird-ass nature, and this chapter is no different. Also, I'm a biologist and am strongly biased toward this field since it's clearly the best and most important area of research. We at Nerd Nite feature so many organismal (and orgasmical) oddities that we had to disperse some of them throughout the book, but we also wanted to bring a few of them to you right off the bat. Dolphins that work for the US Navy? Check. A primer on cephalopods? Check. So here we go . . . release the kraken (or perhaps just normally sized squid)!

—Chris

NERD NITE NYC AND NERD NITE LA

CAMEL SPIDERS: The Rumors of My Size Have Been Greatly Exaggerated

by Forest Ray, PhD

The first camel spider I saw failed to impress me. Don't blame the spider, though—my own expectations were set much too high.

Iraq hadn't yet fallen when my own deployment orders came, but tales of eight-legged behemoths whose jagged mandibles could turn flesh to pudding were already making the rounds, giving us new reasons to "stay alert, stay alive."

That first camel spider skittered through the light of my flashlight while I was on guard duty in Kuwait, as we massed to cross the border. The soldier I stood guard with and I followed it with our lights to get a better view. Sure enough, we had encountered our very first camel spider. It was . . . small.

Camel spiders range in length from roughly two to six inches, with most species closer to the lower end of the range. Those found in the Middle East, however, often grow to between five and six inches in length. Not exactly the stuff of monster movies, but speaking as someone who has been woken up by a tarantula on his chest, I can assure you that terror comes in all sizes.

Stories told of camel spiders certainly do their best to inspire terror, starting with their name. The story—its dominant variation, anyway—is that they earned their association with camels by latching onto the dromedaries and eating them from the underside. Some stories grant the spiders the decency of waiting for the camels to die first, others don't.

Like any great attempt at misinformation, these claims can be tested. Even in the absence of a camel spider, the next time you find yourself next to a camel, poke it in the belly to see if it lacks a self-defense mechanism.

The Great Camel Spider Misinformation Campaign begins, in fact, with the creatures' name. Camel spiders are not spiders at all. They belong to an order of animals called Solifugae, Latin for "fleeing the sun." Other common names include wind scorpions and sun spiders, and it should be noted that they are also not scorpions.

They are, however, to paraphrase Eleanor Shellstrop in *The Good Place*, some ugly motherforkers, at least by humanity's questionable standards.

Perhaps this is why, in the time-honored human tradition of mocking those unlucky enough to have not been born beautiful, so many false stories about them exist.

Beginning with their misleading name, they are said to kill and eat camels, to grow as long as a human forearm, to run at speeds of up to 25 miles an hour and jump up to six feet. Even if you outrun them, they can haunt your sleep, where their venomous bite numbs your flesh, which they strip from your bones. Whatever flesh they don't tear from your body might fall off later, anyway, due to their conveniently multifunctional venom. Oh, and they may also feel inclined to lay some eggs in you for good measure.

We can all feel fortunate that we face no danger of being eaten alive by these little guys, or of becoming unwilling surrogate parents.

In fact, camel spiders aren't venomous and are unlikely to even attempt to bite a human, much as most people wouldn't just stick a pocketknife into something 100 times their size. I mean, sure, there's always one. As my drill sergeant once said before a field exercise: "I don't want to see any of you white boys fucking with the wildlife! Seriously, why do you do that?"

I don't know, Drill Sergeant. I honestly don't know . . .

Although we need not fear being excarnated by a camel spider, it's easy to see where that particular myth comes from. Their mandibles consist of a double set of pincers called chelicerae, which look like the result of a crab bumping uglies with the Predator.

Extending outward from their chelicerae are two appendages, called pedipalps, that look like a bonus set of extra-long legs. These are covered in coarse, sticky hair that helps the camel spider grab its prey and bring it to the wood chipper of its chelicerae.

Camel spiders also use their chelicerae for both survival and mating. Simultaneously. This one is not a myth.

After traveling long distances—camel spiders are quite solitary—to find a female, a male camel spider will creep up on its unwitting potential mate and delicately "massage" her with his pedipalps in an effort to induce a catatonic state of torpor. So that she doesn't kill him. While this sounds objectively terrible by human standards, it gets much worse.

In order to seal the deal, the male needs to get his ball through the net, so to speak, without losing focus on the ongoing life-or-death massage. And due to some questionably intelligent design on the part of camel spider anatomy, this involves one of several complicated maneuvers.

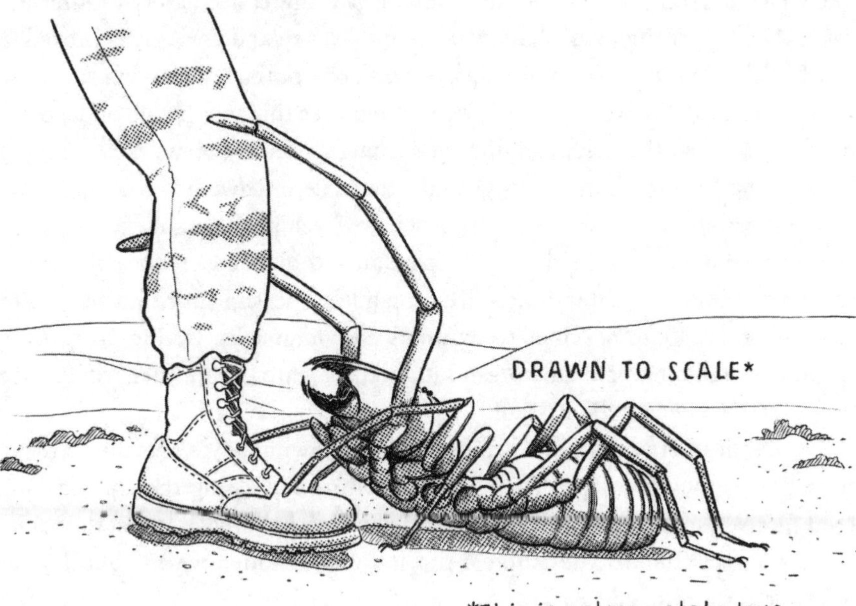

In one, he uses his chelicerae to position the female, then gets his sperm on them and uses them to transfer said sperm into her genital opening. In another, the male first uses his chelicerae to "chew" or massage the female's abdomen and move away from her as she begins to exit torpor. Before she fully regains her senses (assuming all goes according to plan), he releases a capsule of sperm called a spermatophore, which he grabs with his chelicerae and plunges into the female's genital opening. Then he runs like hell.

There is actually a genus within the Solifugae—the *Eremobates*—that simply transfers the spermatophore directly from the genital opening of the male to the genital opening of the female. This is much less fun to describe.

For the truly deviant, the female camel spider's response to the male's pedipalps may be a purely mechanical reaction, as the same behavior has been witnessed in response to being handled in field and laboratory settings. You pervs.

Other myths revolve around the camel spider's physical capabilities. Legend has it that they can run up to 25 miles per hour and jump three to six feet into the air, and that they can and will aggressively chase humans down.

To be fair, camel spiders can hustle at up to 16 miles per hour, but they don't jump much at all. That won't stop them from chasing you.

If you find a camel spider hiding in your shadow and run, however, they will likely give pursuit. This is not out of any innate aggressiveness but rather because they were using your shadow to escape the sun and you rudely moved it.

I fear I'm coming off as a bit of a downer—one of those annoying scientists who steals away the world's mysteries and replaces them with cold and uncaring facts. I'll end, then, with an unsolved camel spider mystery.

As scary as some humans might find these desert-dwelling arachnids, we have almost nothing to worry about. Ants, however, are not so lucky.

Camel spiders will sometimes roll up on an anthill, massacre the colony, and tear up the nest. The marauding solifugid can tear apart an entire ant army with ease, while totally impervious to its prey's counterattacks. Giving no quarter, the miniature avatar of Shelob presses its offensive into the anthill itself, tearing up the earth in its frenzied assault.

For all the energy that camel spiders spend mowing down ants and sacking their homes, however, they have not been witnessed eating the ants, leaving their motive for the attacks an enigma. One leading hypothesis is that the camel spiders burrow into the nests to eat the ant larvae hidden within, but this remains so far unproven.

This murderous habit of the camel spider lingers as a mystery to be solved by those too curious not to fuck with the wildlife.

Forest Ray is a former paratrooper and current science reporter based in Long Beach, California.

NERD NITE MIAMI

MILITARY MARINE MAMMALS: Dolphins So Smart They Should Give Their Own Nerd Nite Presentation

by Laura Chaibongsai

Most scientists would tell you that humans are the smartest animal in the world. Our large brain size, number and density of neurons, and ability to both communicate and use tools place us at the top of the intelligence charts. However, I live in Florida and have personally met "Florida man" so I have my doubts.

Bottlenose dolphins on the other hand are also extremely intelligent and don't throw alligators through drive-thru windows or try to stop hurricanes by shooting at them (#FloridaMan).

Dolphins have a similar relative brain size (encephalization quotient) as humans, they communicate with whistles and clicks, and they have the ability to learn, apply knowledge, and even recognize themselves. They can alter their surroundings and use tools; some dolphins will cover their rostrum with a sponge to protect themselves as they dig up prey from the ocean floor. And most important, many believe dolphins have sex for fun, so obviously they're smart.

In addition to their brilliance and trainability, bottlenose dolphins can dive to nearly 1,000 feet and have sophisticated natural sonar, so they make excellent partners for the US Navy. Since 1959, the US Navy has been training dolphins to detect underwater mines and locate "unauthorized" swimmers or divers near their harbors in the United States and abroad. These highly trained service members mark a mine or attach a device to the diver and return to their boat unharmed for a fishy treat.

Navy seals, well, actually, sea lions, are also an integral part of the navy's Marine Mammal Program. California sea lions have the ability to dive deep, can maneuver through tight, cluttered areas like harbors, and have excellent vision and hearing. The navy has relied on them to recover equipment that is dropped into the ocean—the navy's gear isn't cheap, so they want it back!—as well as patrol harbors like dolphins do.

In addition to their service to the military, navy dolphins and sea lions have been the focus of over 1,200 scientific publications throughout 60-plus years of research, ranging from studies on sleeping behavior to the impacts of oil exposure after the 2010 Gulf of Mexico oil spill, contributing greatly to our understanding of both captive and wild marine mammals.

While someday soon robots or drones will take over most of their tasks, the technology may never match the exceptional skills of the dolphins and sea lions who have supported the US military for over sixty years.

Laura Chaibongsai is a marine science nerd and the boss of Nerd Nite Miami. As a professional science communicator, Laura specializes in citizen science, community engagement, research development, and finding creative ways to translate ridiculously complex research topics for nerds of all ages and backgrounds!

NERD NITE MADISON

SEX CATAPULTS!

by **Ben Taylor**

Catapults: They're amazing! Stored potential energy rapidly released as forward motion to launch missiles, basketballs, watermelons, and even flying cars through the air. We know them, we build them, we love them. But humans are not alone in our adoration of the contraption, for while nature may abhor a vacuum, she loves a nasty catapult. And nature puts catapults in one fun place in particular—namely, the bodies of arthropods.

Arthropods, from Greek for "jointed foot," are a group of animals made up of insects, crustaceans, spiders, and the various creepy-crawlies of the planet. The jointed limbs in the hard chitinous exoskeletons of arthropods act as ideal spots for springs, locks, and hydraulic mechanisms that release stored energy in a near instant when triggered. Arthropods are catapult creatures.

Take, for example, the extreme release of stored power in the bite of the trap-jaw ants. Hundreds of different ant species have evolved a mechanism in their mouths similar to a bow and arrow, which latches open their mouthparts when they're awaiting prey or preparing to escape a threat. When triggered, the mandibles snap shut at speeds up to 2,300 times faster than a blink of a human eye. This allows for the instantaneous snagging of food, or in some cases an evasive maneuver where the ant releases its jaws against the ground to launch itself through air and away from danger.

Most arthropodal catapults, called modes of kinematic transmission, have evolved as a means of both predation and defense. But in April 2022, a paper published in *Current Biology* changed everything. Scientists at Hubei University in China, writing about the male hackled orb weaver spider, described a behavior that very much utilizes catapults for sex. Why does the male hackled orb weaver care about catapults during sex?

To start, the female hackled orb weaver spider is *way* larger than him—nine times larger. As if that didn't make sex intimidating enough, the female hackled orb weaver spider engages in post-coital snacking, mainly on her male partner. This George Costanza–esque sexual cannibalism is seen in many size-dimorphic pairings, perhaps most famously with the praying mantis (for with praying mantises, the male is always giving head), and is an extremely efficient way of getting extra protein for egg production. It's just not super ideal for the male if he wishes to live and mate another day. So this hackled orb weaver spider has evolved a very special catapult, on a very special part of his body. And we're not talking about boner stuff. We're talking hydraulics, another catapult-friendly aspect of the bizarre and beautiful bodies of arthropods.

Instead of typical veins and capillaries, arthropods push blood and fluids through their system using a massive tubular heart that stretches the length of their bodies. Imagine squeezing a condom full of blood. Go on. Imagine it. When you squeeze that extra-large condom full of red, red blood, you force fluid to rocket to the parts of the condom outside the zone of your squeeze. You force those portions of the condom to quickly expand with the force of the sudden hydraulic pressure of your squeeze. You have made a liquid catapult of sorts, and this is how nature makes it happen with male hackled orb weaver spiders.

This catapult is on the spider's skinny front legs, each thinner than a human hair. During copulation, the male hackled orb weaver bends its jointed legs against the mounted female, as though preparing a tiny push-up against her. When mating is over and mealtime set to begin for the female, powerful muscles in the male spider's thorax smash body fluids into its folded forelimbs, and the resulting hydraulic force catapults the spider off the female at speeds up to three feet per second. The male lives to mate another day, and scientists even observed one male catapulting his way out of bed six times in eight hours. Not bad, little male hackled orb weaver spider.

And that is the beauty of arthropods. All that potential hydraulic energy, just sloshing around: might as well force it into a forelimb so you can escape from bed without having to be the post-sex room service as well.

Ben learned a love for insects growing up in the mountains of Albuquerque, New Mexico. He received his degree in entomology from the University of Wisconsin–Madison and spent the following year chasing native bees through Wisconsin apple orchards. He has been stung in the face. Ben has served as a youth programs coordinator at the National Museum of Natural History and is currently the education manager at the Horticultural Society of New York. He has been the boss of Nerd Nite Madison and Nerd Nite DC and currently guest-hosts Nerd Nite NYC.

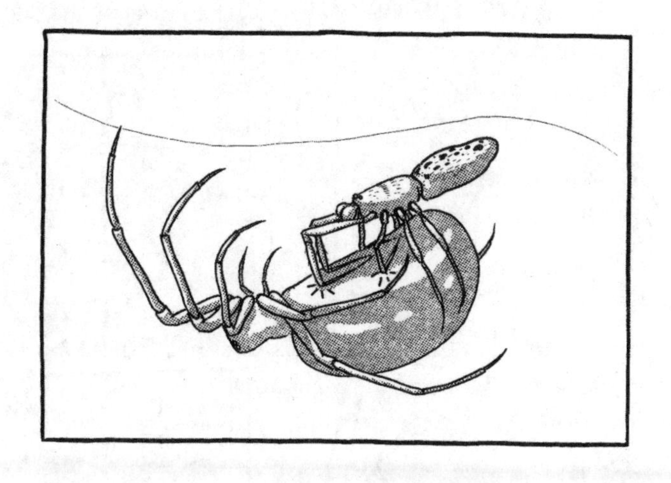

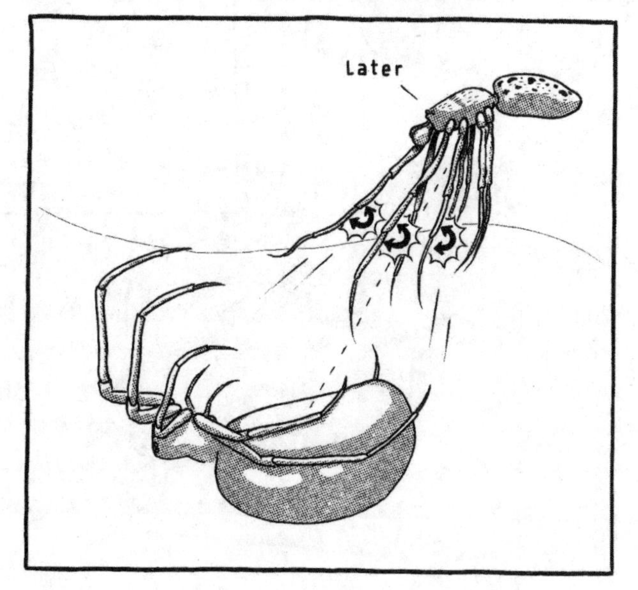

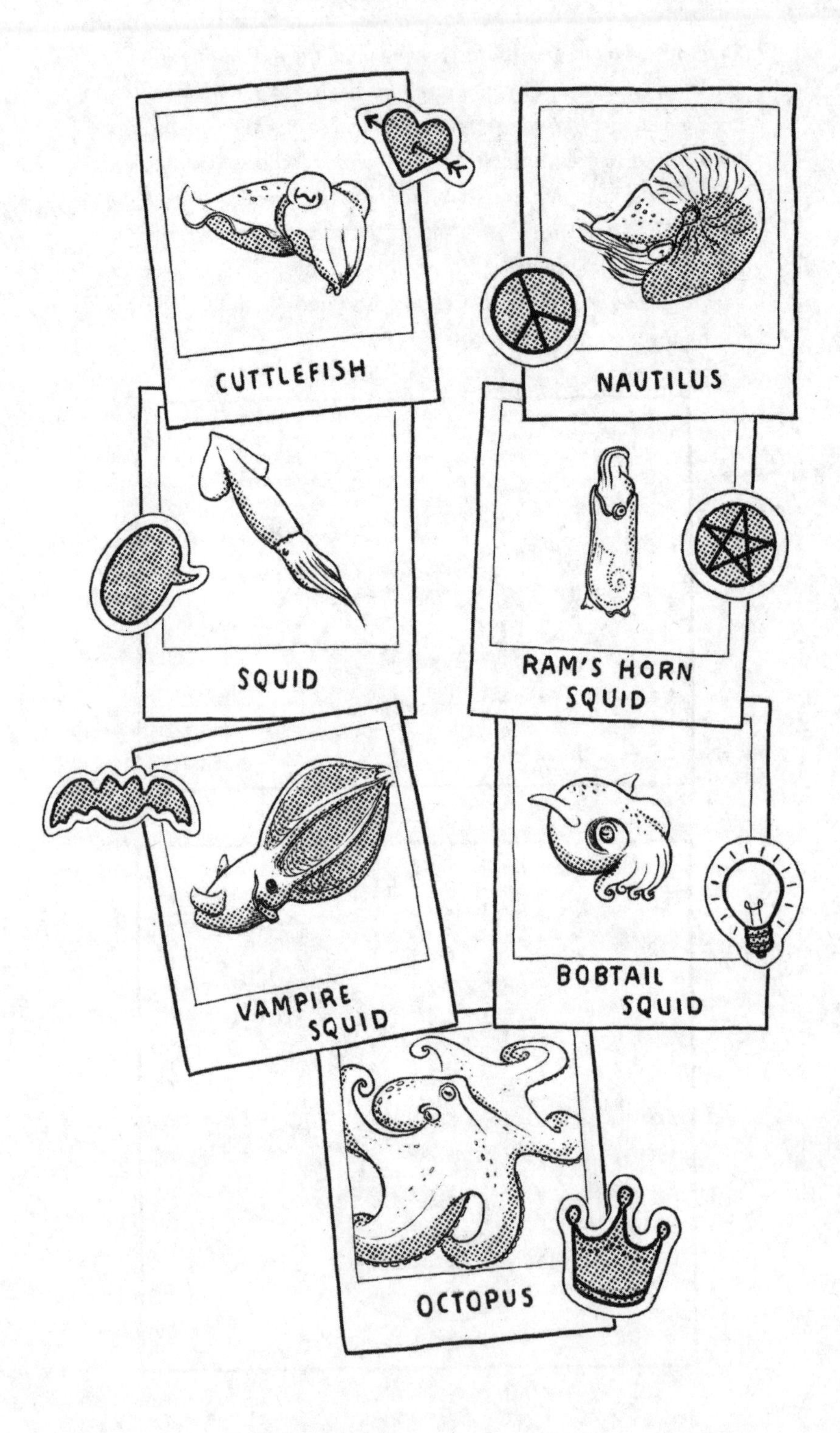
CUTTLEFISH
NAUTILUS
SQUID
RAM'S HORN
SQUID
VAMPIRE
SQUID
BOBTAIL
SQUID
OCTOPUS

NERD NITE MELBOURNE

CEPHALOPODS: The Impossibly Awesome Invertebrates

by Aerie Shore

Cephalopods—denizens of kingdom Animalia, phylum Mollusca, class Cephalopoda—are the most intelligent and behaviorally sophisticated of the invertebrates. Their ancestral lineage surfaced a mind-blowing 500 million years ago, and that generous girth of evolutionary time has endowed them with anatomical superpowers and thrilling attributes. Come with me on a cephalopodyssey, as we briefly visit the seven presently acknowledged extant orders of these magnificent underwater weirdos.

Order Sepiida: CUTTLEFISH

~120 KNOWN EXTANT SPECIES

Diving in hypothermia-inducing winter waters near Whyalla, South Australia, I had the profound pleasure of witnessing the mating rituals of giant Australian cuttlefish. Larger males compete for mating rights in flamboyant, Technicolor displays of mightiness. Males too small to contend in these showdowns of cephalopodic machismo utilize a clever tactic, earning the nickname "sneaker males." Sneakers tuck in their telltale, sexually dimorphic arms and disguise their coloration to mimic female appearance. Such subterfuge enables them to surreptitiously sidle up to a potential paramour, reveal their true sex, and get it on in a tentacular face-to-face mating embrace. Cuttles have eight arms and two tentacles. If you think making the beast with two backs is boastworthy, try making the beast with 20 appendages. Think naked underwater Twister with four of your more adventurous friends. For science!

Order Nautilida: NAUTILUS

3–6 KNOWN EXTANT SPECIES

Our pelagic friend sports a bombastically bearded visage of slender cirri and a countershaded, mineralized external shell composed of multiple camerae (chambers) formed in spiraling succession as it matures. The squishy body of this mustachioed mollusk inhabits the outermost camera and controls buoyancy via the siphuncle, a conduit running through all camerae to regulate the gas:liquid ratio within. Is your parent's brother secretly a nautilus? Sweet! You've got a funky uncle with a siphuncle . . . *and* a radula, the rasping tongue unique to mollusks.

Order Teuthida: SQUID

~300 KNOWN EXTANT SPECIES

Squid possess eight arms, two tentacles, cooperative hunting prowess, and boisterous visual communication. Some species display patterns on their exterior surfaces for inter-squid-squad signaling, created with chromatophores, iridophores, leucophores, and photophores. For some, specific arm positions also bear significance. Male squid have been observed bifurcating their mantle pattern such that one side boasts an alluring display to entice the ladies while the other broadcasts intimidation and discouragement to rivals. It's certain that specific patterns and postures convey meaning, yet the Rosetta Stone of squid semaphore remains elusive. Earth's most massive invertebrate, keeper of the largest eye in the animal kingdom housed in the ammonia-rich temple of its magnificent body, is the colossal squid.

Order Spirulida: RAM'S HORN SQUID

1 KNOWN EXTANT SPECIES

Cephalopods rising up from the shadowy abyss, adorned with two spiraling horns like a Balrog . . . that's what I fiendishly envisioned while collecting the namesake mineralized shells of *Spirula* on the coast of New Zealand. Alas, these dolma-sized creatures form but a single shell *inside* their body. It facilitates buoyancy control, not ancient magic steeped in shadow and flame. More closely related to extinct belemnites than to squid, *S. spirula* are too seldom observed to divulge juicy details of their deep-sea lifestyle. They brandish eight arms, two tentacles, and a single green photophore at mantle tip but lack a radula.

Order Vampyromorphida: VAMPIRE SQUID

1 KNOWN EXTANT SPECIES

Vampyroteuthis infernalis, the endearing vampire squid from hell, has a gelatinous body and webbed arms lined with spooky-looking cirri. Notable items in its squishy bag of tricks include ejectable, bioluminescent mucus enabling it to escape bedazzled predators, an ability to survive and thrive in extremely low oxygen zones, and photophores along its body that produce light with, yes!, the enzyme luciferase. Like true squid, it harbors an internal gladius, a hard internal bodypart shaped like a sword. Like octopi, it wields eight arms and zero tentacles. What differentiates the vampire squid from other taxa are two retractile filaments (for sensing disturbances in the Force) and that it never drinks . . . wine. While most cephalopods are semelparous, reproducing only once near the end of short life spans, *V. infernalis* flouts the tradition of "mate once, die young" by spawning multiple times over a heroic ten-year life span.

Order Sepiolida: BOBTAIL SQUID

~70 KNOWN EXTANT SPECIES

In a brilliant display of *squid pro quo*, Hawaiian bobtails engage in mutualistic symbiosis with the bacterium *Vibrio fischeri* to emit bioluminescent light. The nocturnal bobtail modulates this light and projects it downward, matching the intensity of moonlight penetrating ocean water above, thereby evading death from below. Bobtails reap the benefit of counter-illumination camouflage, and *V. fischeri* get reliable nourishment.

Order Octopoda: OCTOPUS

~300 KNOWN EXTANT SPECIES

Which features of the octopus are spectacularly awesome? ALL OF THEM. Octopuses are the souped-up hot rods of the animal kingdom. They have three hearts circulating blue, hemocyanin-rich blood; neurons distributed throughout their absurdly contortable bodies; suckers that taste; skin that "sees"; distinct personalities with a spectrum of moods; a savvy penchant for fish-punching; and the cognition to use tools, solve puzzles, and form lasting opinions about you. Despite color-blind eyes, they easily master physical and visual camouflage. Researchers propose that octopuses perceive properties of light with cells in their skin. They readily release ink in defensive smoke screens, and another brilliant predator-evasion tactic is to mix ink with crucial amounts of mucus to squirt out a pseudomorph—a sacrificial decoy. Some clever octopodes eject a row of pseudomorphs and, matching body color to ink blobs, hover next to them . . . hiding in plain sight!

Octopus intelligence is vaguely equated to that of a cat or a fifth grader, although I have yet to see either of those squirt a passable pseudomorph out of its butt or open a jar *from the inside.* By that reckoning, eating an octopus is the criminal equivalent of feeding your cat to a fifth grader . . . and then eating that fifth grader. These shape-shifting ninjas of unquantifiable awesomeness deserve to flourish in the ocean, not perish on your plate.

Aerie Shore is an independent science communicator, kink educator, accordionist, ceramic artist, and creator of tentacular cephalopottery. Studies in marine and animal biology whisked her from California to New Zealand and ultimately on to Australia, where she resides near the ocean on the Surf Coast.

NERD NITE DC

STOMATOPODS: Why Is My Thumb Bleeding and My Mind Blown?

by Peter C. Thompson, PhD

There is a creature lurking out there in the ocean depths that you really need to know a little bit about. You might think by looking at them that they're shrimp or lobsters, but they're a unique group of creatures called stomatopods that seem underappreciated in the popular marine biology you encounter in wildlife specials. They have some truly amazing adaptations, claiming title to both the fastest strike in the animal kingdom and the most complex eyes. For these characteristics and more, let us pledge a little respect to mantis shrimp, the cursed thumbsplitters of the sea.

Mantis shrimp is the common name for a group of crustaceans called stomatopods. These active marine creatures range in size from less than half an inch to more than a foot long. They sport "arms" that are highly specialized weapons resembling those of a praying mantis. Technically, this weapon is termed the raptorial appendage, and though it would be inappropriate to call it a claw, it resides in the same position as the claws on a lobster. Most species live in subtropical climates. Many have only been observed in a small local area, and consequently they are at high risk of extinction due to climate change and cascading effects of human activities like fishing and aquaculture.

Stomatopods may not have gotten much attention because they are often tough to spot in the wild thanks to fantastic camouflage. They're quite adept at hiding in their community, whether that be a coral reef, a sandy bottom, or the muddy bottom of an estuary. The most diverse (and colorful) stomatopod species occupy burrows and crevices in coral reefs. Typically, they only show their stone-gray parts, but when threatened or courting a lady stomatopod, they can reveal vibrant pinks, yellows, and greens. In between the coral reefs, ghostly white species live in complex burrows in sand composed of broken coral skeletons. A few inconspicuous green and brown species dig lairs on muddy bottoms and prey on unsuspecting passersby. Though difficult to find, when these creatures decide to attack, the results are hard to ignore.

It is not an understatement to say that the attack of a mantis shrimp is mind-blowing. Large stomatopods can deliver a blow with the strength of a *.22-caliber bullet.* They do this despite being submerged in water that hinders their every movement. Their attack can crack the shells of crabs and snails, capture passing fish and shrimp, and knock the arms off rivals as they defend their homes. When caught in a fishing net, they may lash out at their captors, causing extreme damage to any soft body part offered. This has earned them the lovable nickname "thumbsplitters" among fishermen who curse them as pests.

Stomatopods can be divided into the spearers and the smashers based on the shape at the end of their respective weapons. Spearers have a front appendage with tines like a pitchfork; when propelled at blinding speeds, these can pass completely through a fish (or a thumb). Smashers, on the other hand, prefer to hammer on their prey using a clublike weapon. Their strikes have been known to crack aquarium glass when a feisty stomatopod decided to pick a fight with its reflection in a tank wall. Given the remarkable nature of this strike, it is worthwhile spending a little time trying to put the speed and force into a bit of perspective.

A University of California–Berkeley group engaged a PBS crew with a collection of high-speed motion cameras and force sensors to try to understand the phenomenon. To adequately capture the speed and force of the strike, they had to employ a camera capable of 20,000 frames per second—for comparison's sake, most films we watch are shot at 30 frames per second. The acceleration of the tip of the raptorial appendage was found to be 100,000 meters per second. In more familiar terms, that's 0 to 60 in less than a second. The force calculations were equally perplexing: The force of the initial strike was upward of 2,000 Newtons. This is the equivalent of a 225-pound person stepping on your toe in heels. While these statistics are amazing, the story gets more interesting when you look at the mechanics of it all.

The strike is a remarkable feat of evolutionary ingenuity in energy storage and force amplification. The strike is powered by typical muscles that are not excessively large. The secret is a series of catch mechanisms, similar to a ratchet, that allow ordinary muscles to store energy in a flexible piece of shell until the stomatopod decides to release it. A "click" mechanism engages to hold the striking appendage in place while energy loads into a "saddle" that flexes in two directions, allowing it to store more energy than would seem likely. Another click mechanism provides support for more force loading into the saddle, similar to a classic windup toy.

When the stomatopod decides the time is right, the catches are released simultaneously, freeing the stored energy in the saddle in less than three milliseconds. Upon release, the weapon expands from three distinct joints to its maximum length. Collapsing joints extend the blow farther out than would be possible on a simple rotating arm, thereby increasing the impact of the attack. The movement of the weapon is so fast that it literally *rips apart the surrounding*

water, forming small bubbles. The bubbles burst while in contact with the target, releasing another round of energy with a force that equals approximately half the initial impact. This process is called cavitation, and it's also notorious for causing metal fatigue in underwater propellers. There are few organisms with the protection to withstand this one-two punch, all generated by a creature that you could hold in your hand.

If their insanely strong strike wasn't enough to pique your interest, stomatopods hold another record: Their eyes are the most complex in the animal kingdom.

These eyes are compound, similar to those of insects. Rather than a single lens, the eye is built from many small lenses that form an image by combining multiple smaller images. What sets stomatopod eyes apart from others is that each eye is composed of two distinct hemispheres, separated by a specialized midband. This unique arrangement allows a stomatopod to form three separate images in each eye, providing remarkably acute depth perception for a relatively narrow field of vision. Additionally, the midband may have as many as 12 different color-sensing pigments. This is the highest number recorded in any animal and three times what is found in humans. This should provide stomatopods with amazing color discrimination, even extending into the incredibly broad range of ultraviolet colors. Though we can only speculate, these eyes may be vital for controlling their deadly attacks or, more likely, be important for observing their surroundings and communicating with other marine organisms.

So . . . the next time you are on a dive (or maybe just in an endless YouTube session) you should keep your boring human eyes open for these remarkable little creatures. There are likely several hundred species of stomatopods waiting to be discovered. Who knows what adaptations they may have that might shock and impress? Like tearing water!

Pete is an evolutionary biologist who studies parasites for a living. He has managed to convince the federal government that he cares about cattle, but his heart remains in the sea with all its amazing creatures. Also, his heart isn't literally in the sea.

NERD NITE OKINAWA

FINDING NEMO('S SEX): Sex Change and Gender Roles in Anemonefishes

by Jann Zwahlen

The words *gender* and *sex* are often used interchangeably, but there are some differences between them. Sex is the trait determining whether male or female gametes are produced in a sexually reproducing organism, while gender is a range of characteristics related to masculinity and femininity and normally used only in a human context. Here, both terms are used due to the human-centric approach of this contribution.

Who hasn't heard of Nemo—one of the most famous fish in pop culture history with its iconic look? Nemo is a clownfish, and clownfish are among 28 species of anemonefish distributed in the Indo-Pacific. As the name suggests, anemonefish live in symbiosis with giant sea anemones, which basically form their home for their entire lives. Luckily, the fish don't live alone, but in a colony that, at first blush, resembles a monogamous couple with some immature juveniles. But this seemingly classic 1950s "family" didn't start out that way. After being deposited as eggs on a hard substrate, hundreds of larvae hatch and move into the open ocean; from there, they will return to reefs approximately two weeks after hatching. And this is where the difficult part begins. These juveniles aren't returning to the anemone where their biological parents live, but rather they're simply looking for any anemone that is accepting residents. And when they succeed in finding a welcoming colony, they can then begin life as immature juveniles. But such a life isn't always easy, as they'll spend time defending their rank within the colony while they're waiting . . . waiting for an adult to die.

And it's very important to note that the juvenile fish at this point in their lives are intersex—they contain immature gonads of both sexes.

Each colony is strictly hierarchical, and rank-based, with the largest individual, rank 1, being the female, followed by the male in rank 2, and then the juveniles in ranks 3 and higher. The death of a high-ranking individual opens a window for lower ranks to move up one rung and eventually become male by maturing their male gonads. To finalize their life cycle, the males wait for a female to die, so that they can take over her position. In that case, the female gonads mature, while the male parts disappear. This gonadal sex change is also associated with other changes, such as body size, coloration, and behavior.

Knowing that their life history is rather exceptional, one might wonder if at least in day-to-day life, anemonefish colonies resemble the "idyllic" and "traditional" (stereotypical) human family structure with the man providing and protecting, and the woman taking care of the offspring. Well, of course not. The large female dominates the whole colony and is the major aggressor toward other fish to protect the colony. The male, on the other hand, is submissive to the female and takes care of most reproduction-related tasks, such as cleaning the substrate for the eggs and aerating the eggs regularly using his fins. Thus, the gender roles definitely don't follow the "traditional" human family, and life in a colony is rarely idyllic—in some anemonefish species, the colony members seem to fight constantly.

Finding Nemo is often deemed surprisingly accurate, but is it really?

Starting with the parents, the father Marlin is bigger than the mother, Coral, and Marlin is also the one who tried to fight off the barracuda—so exactly the opposite of what we'd expect of real-world anemonefish. Even worse, Marlin is afraid to approach the eggs, but in the sea that would mean almost certain death for all of them due to lack of oxygen—a bit inconvenient. Now, if we ignore the major flaw that Nemo didn't have a larval phase, there are still several problems associated with him. Nemo is male at hatching, but actually he should be intersex until he can climb to rank 2. "Luckily" for Nemo, the rare opportunity to climb a rank presents itself with the death of his mother, Coral. Especially in the period between *Finding Nemo* and its sequel, *Finding Dory*, Nemo and Marlin should have climbed one rank and changed their sex to male and female, respectively. Therefore, according to the anemonefish life cycle, Marlin and Nemo should form a reproductive couple in the sequel. Frankly, this thought is quite disturbing and shows beautifully the importance of a larval stage in these iconic fish.

Thus, though it was hugely successful at the box office, *Finding Nemo* strikes an awkward balance between being accurate in some situations and wildly inaccurate in others. Of course, it is ultimately fiction, but it missed out on a perfect opportunity to introduce a nontraditional family model into the mainstream simply by following the anemonefish biology a bit more carefully.

Jann Zwahlen is a marine biologist currently doing his PhD on anemonefish acclimation and adaptation in Okinawa, Japan.

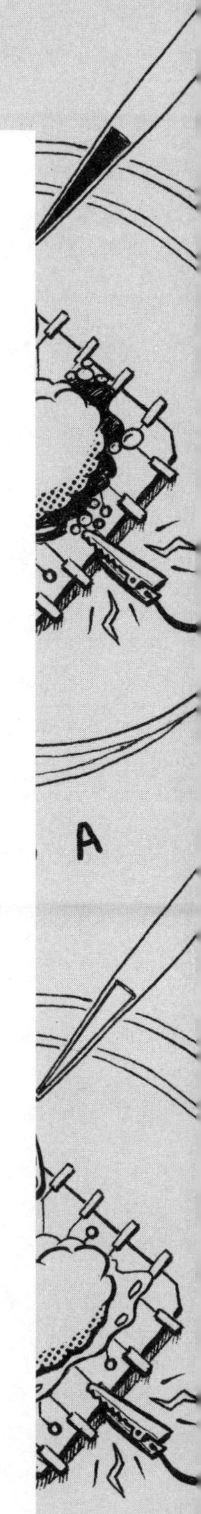

Mmm . . . Brains

Please let us get this out of the way. The contributions in this chapter are *not* about zombies. We're sorry. Though do consider this a necessary public service. But after *The Last of Us*, *The Walking Dead*, *Fear the Walking Dead*, *The Talking Dead*, *Zombieland*, *World War Z*, *Juan of the Dead*, *Shaun of the Dead*, *Night of the Living Dead*, *The Return of the Living Dead*, *The Day of the Dead*, *Zombie Flesh-Eaters*, *Let Sleeping Corpses Lie*, *Little Monsters*, *28 Weeks Later*, *Dead Alive*, *Night of the Comet*, *I Walked with a Zombie*, *Night of the Creeps*, *Thriller*, *Re-Animator*, *Rec*, *Warm Bodies*, *One Cut of the Dead*, *Versus*, *Death Dream*, *Cemetery Man*, *Train to Busan*, *Scooby-Doo on Zombie Island*, and—dear God, soooo many zombie movies and shows—you're probably sick of them by now. Though based on the girth of the aforementioned list, maybe your appetite for zombie flicks remains as insatiable as any zombie's appetite for our tasty gray matter? Nonetheless, instead of the undead, psychology and neuroscience are about to take center stage in the next few pages, where you'll encounter tales about hearing colors and tasting sounds, as well as what mental gymnastics your brain must do to speak without an accent. You'll also learn why we humans are so gross. And, of course, you'll understand how ethics, pleasurable sex, and happiness all come from within our supple, succulent, fresh, locally sourced, free-range, not-entirely-GMO-free, delicious . . . mmm . . . brains.

—Matt

NERD NITE LONDON

IT'S NOT YOU, IT'S MISOPHONIA

by **Dr. Jane Gregory**

I am three years into a project researching why I have the urge to kill my husband when he eats a packet of crisps in the same room as me. That's not a joke. After a decade of working as a clinical psychologist by day and listening to my husband's Darth Vader breathing by night, I decided to go back into academia to study the phenomenon of misophonia, when small repetitive sounds cause big negative reactions.

Misophonia can attach itself to any repetitive sound, but the most common ones are things like chewing, breathing, sniffing, and throat clearing. It can be hard for sufferers to talk about because of how difficult it can be to tell someone politely that the sound of them keeping themselves alive is repulsive to you.

Never heard of misophonia? You're not alone. Only 14 percent of the general population in the UK have heard the term before, and only 11 percent of the population in the US are familiar with it (according to "Misophonia in the UK: Prevalence and Norms for the S-Five in a UK Representative Sample" and "Public Awareness of Misophonia in U.S. Adults: A Population-Based Study" in *Current Psychology*, respectively).

Think you might have misophonia? You're also not alone. Almost one in five people have misophonia to the point where it causes problems in their lives. That's right, there are more people who have it than there are people who have heard of it.

I learned about it from an article in *The New York Times* that was sent to me by several of my loved ones. Seemingly from anyone I had ever glared at for rustling a packet of sweets next to me. Interesting that they chose to notify me instead of using this opportunity to look at their own behavior. It's not that hard to transfer your treats into a silent cloth bag in private before you leave the house, people!

It was a relief to discover a name for this after spending my entire childhood believing I was just snobby and intolerant. I wish I could visit myself as a teenager and tell her, "It's not you, it's misophonia! Well, it's a little bit you. But don't worry, being the right amount of snobby and intolerant will also make you welcome at Oxford in 20 years' time."

Whenever I talk about misophonia—which is often—at least one person lights up with the realization that they are not a crazy person after all. And at least one other person insists that it's not a real thing because no one likes the sound of loud eating. Fortunately, there's a stat for that. Actually, hypothetical naysayer, only 85 percent of people don't like the sound of loud eating. So there.

So, what's the difference between not liking a sound and having misophonia? Misophonia is more complex than simply not liking a sound. When developing a questionnaire to measure misophonia, my colleagues (shout-outs to Silia Vitoratou, Nora Uglik-Marucha, and Chloe Hayes) and I discovered there are five key aspects of the misophonia experience:

1. A feeling of emotional threat, like feeling trapped, panicked, or helpless if you can't get away from certain sounds. For some, it's a fight-or-flight-type response, which is the technical term for the surge of adrenaline you get when you run away to avoid slapping someone for the way they slurp their tea.

2. A sense that there is something wrong with you for the way you react to sounds.

3. A sense that there is something wrong with the other person for making the sounds. As the old saying goes, *Sounds don't kill people. People who make disgusting sounds make people want to kill people.*

4. Being aggressive to others who make annoying sounds or worrying that you will act out one day. It's far more common in misophonia to worry about losing your temper than it is to act on that temper.

5. Feeling as if you are missing out on things because of the way you react to sounds.

It's helpful that our brain can tune into subtle sounds that could be signs of trouble. We all have the capacity to do this in times of potential danger. One night I was walking home in the wee hours and my brain tuned into faint footsteps, and even though there were competing sounds of wind and traffic, my brain wouldn't let the footsteps go. As the sound got closer, I started running. I then heard a loud shout, "Don't worry! I'm not going to attack you!" Which felt like the sort of thing a psychopath would probably say and I certainly didn't slow down to find out.

My misophonia will also serve me well when the zombies arrive. You're going to want me on your team. While you're busy hoarding weapons, food, and mobile phones, I will be on guard for the sound of scuffing feet and heavy breathing. I will be ready to protect you. But if you eat that food too close to me or use that phone without headphones, you're on your own.

If you're thinking this frivolous topic is not a worthy use of precious research funds, then let me reassure you: Many, many funding bodies agree with you. I was turned down for six small project grants before finally being awarded Wellcome Trust funding to focus on misophonia as a clinical researcher at Oxford. Getting that level of funding after all those rejections was like auditioning for a role as an extra on six consecutive Bond films, and then a major medical research funder calls me up to offer me a date with Daniel Craig. I just hope he keeps his mouth closed when he eats.

*Dr. Jane Gregory is a clinical psychologist researching misophonia at the University of Oxford. After decades of glaring at strangers for daring to eat crisps in public, she felt vindicated to find out there was an actual name for her special breed of intolerance. She has since dedicated her career to finding more reasons why she is, categorically, not the a**hole here.*

NERD NITE MIAMI

SEX, DRUGS, AND HAPPINESS

by **Melissa Blundell-Osorio**

An introduction to positive psychology. Wait. Is there negative psychology? Well, sort of!

Traditionally, psychology focused on what was wrong with people (conditions such as depression, anxiety, mental illness, et cetera), how to fix it, and how to get people from a deficit to a more neutral place. But positive psychology emerged as a movement to focus on what is right with people and how to build on that, as well as how to get people from neutral to a place where they're thriving and satisfied with life. Just because you're not depressed doesn't mean you're happy, just as not being poor doesn't mean you're rich and not having an illness doesn't mean you're necessarily in great health.

One of the biggest myths about happiness is that it's largely dependent on our circumstances and the things that happen to us, meaning that it's not really within our control. But this is false (obviously, because otherwise I wouldn't have written this).

Researcher Sonja Lyubomirsky has looked at what accounts for the difference in happiness level between any two people. I have a certain level of happiness, that is, but my next-door neighbor has a different level of happiness. What accounts for this difference?

Originally, Lyubomirsky found that about 50 percent of the difference in happiness level between any two people is due to genetics; we all have a biological set point. She estimated that about 10 percent of the difference is due to our circumstances—external factors like age, how much money we make, where we live, and whether we're single or in a relationship—and that up to 40 percent of that difference is based on activities within our control. Since her original study, other scholars have published work suggesting that these numbers might be slightly different. However, the fact that we tend to overestimate the impact of external factors on our life satisfaction and that intentional actions on our part can increase our level of happiness is not disputed.

Our intentional activities include actions and behaviors such as our habits, the things we do, and the things that influence how we think and feel. Our goal is to leverage these activities to the best of our abilities. That's how we can change our happiness set point. Our happiness levels go up and down in response to events that happen in our lives, though they remain relatively stable. They might spike up or down but will eventually return to their original set point.

People adapt over time to various events. For example, newlyweds experience a spike in happiness and then go back to their original happiness set point about two years in. Thus, a change in our lifestyle or habits is required to change our set point so that it's relatively stable at a higher level than before.

One thing to keep in mind is that our brains are not wired to think optimistically. Instead, they're wired with a negativity bias. Anything that's bad is going to be stronger and more attention-getting than something good. This is a function of how we evolved—for our early ancestors, optimism didn't offer an evolutionary advantage. The person who left the cave and was like "Oh, I'm *not* worried about predators" didn't survive for long. Conversely, the one who was on the lookout for threats was more likely to survive, and therefore more likely to pass on genes. Our brains are wired to protect us and to look for anything that could potentially go wrong.

We've likely all had the experience of getting feedback that was maybe 95 percent positive, but we end up focusing on that 5 percent that's negative. We're more likely to stay up at night worrying, rather than finding ourselves unable to sleep because of all the things we're grateful for. For this reason, we have to actively put the positive into our lives, actively leveraging that 40 percent through happiness-building activities like fostering positive interactions with people and focusing on what's right with us instead of what's wrong, because our brains are not naturally going to do this.

Luckily, there are many, many ways we can increase our life satisfaction! This is where the pathways of the PERMA-V model come in:

- Positivity: Building our experience of positive emotions through pleasure, gratitude, curiosity, and the right mindset.
- Engagement: Being present and fully experiencing our lives through mindfulness and flow experiences.
- Relationships: If any of these pathways is more important than the others, it's this one. In study after study, research shows that what the happiest people have in common is that they have strong, authentic, high-quality relationships. Positive psychology can be summed up in three words: *Other people matter!* We're wired for connection; we need people that we care about and that care about us. We need community.
- Meaning: Purpose; what you do needs to matter to you. For many people this is a spiritual pathway, feeling like you're part of something that's larger than yourself.
- Achievement: A big part of flourishing is simply being able to do the things you want to do and accomplishing the things that are important for you. This is where goal setting and motivation come in.
- Vitality: Overall health, sleep, exercise, nutrition. Your mind and body are not separate things; how you care for your body affects how you think and how you feel.

Ideally, all six of these pathways will be high. More realistically, they'll fluctuate throughout your life. At different points some will be higher than others.

Now let's get to the good stuff. The practical stuff. How do money, drugs, sex, and having kids affect our levels of happiness?

- Money: Will it make me happier? Yes, to an extent. The difference in happiness levels between someone who makes $17,000 a year and someone who makes $70,000 a year is dramatic. However, the difference between someone who makes $70,000 and someone who makes $700,000 is not nearly as dramatic. *More money means more happiness until your basic needs are met.* After that, more money doesn't buy more happiness.
- Spending money doesn't bring lasting happiness. It's very easy to fall into the trap of what's known as the hedonic treadmill, which essentially dictates that whatever level of wealth or material goods you have, you quickly adapt and always want more. You may receive a temporary spike in happiness by spending money, but you have to continue

buying more and more things in order to maintain that level of happiness. You're much better off spending money on an experience, particularly an experience you can share with someone else.

- Drugs: Will they make us happier? Depends. It depends on what drugs we're talking about. If you have a heroin habit, your life isn't heading in the right direction. However, a lot of research suggests that hallucinogenic drugs like psilocybin, or magic mushrooms, can have lasting positive effects on mental health. When such drugs are used responsibly, people report having meaningful experiences that feel spiritually profound. They feel more connected to the universe and to other people. The drug can lift their mood and improve their psychological health long-term—even friends and family members report that they're kinder, calmer, and happier.

- Sex: Will it make us happier? Duh. In one study, people ranked sex as the number one most pleasurable activity (which was groundbreaking science that nobody knew). It also ranked number one for meaning and engagement. It's unique in that it fits into many pathways at once. Sex is considered pleasurable, engaging, and a contributor to vitality; depending on who you're with, it can be meaningful and will often contribute to the deepening of a relationship. Shit, throw in a goal to have more meaningful sex and you'll cover all six pathways at once.

- Kids: Will they make us happier? Maybe. This largely depends on what your personal definition of happiness is and what you value. If you define happiness as having fun and engaging in pleasurable activities, kids will likely not make you happier. Raising kids ranks low on the pleasure scale, and parents report lower levels of having fun than do non-parents. What parents do have are very high levels of meaning. They find having kids is incredibly rewarding; their children offer them an opportunity to transcend, as life is about more than just themselves. If you define happiness as engaging with something meaningful, then having kids will likely make you happier.

Here are some practical next steps you can do to work toward continued happiness. You can:

- Keep a gratitude journal.

- Perform random acts of kindness.

- Increase flow experiences.
- Set a goal and work toward achieving it.
- Meditate.
- Engage in physical activities.
- Shorten your commute.
- Spend more time in nature.
- Deepen your relationships.
- Get involved with a cause you care about.

Variety is key!

Melissa Blundell-Osorio was a Nerd Nite Miami boss from 2014 to 2021. She holds an MSc in Human Evolution and Behavior.

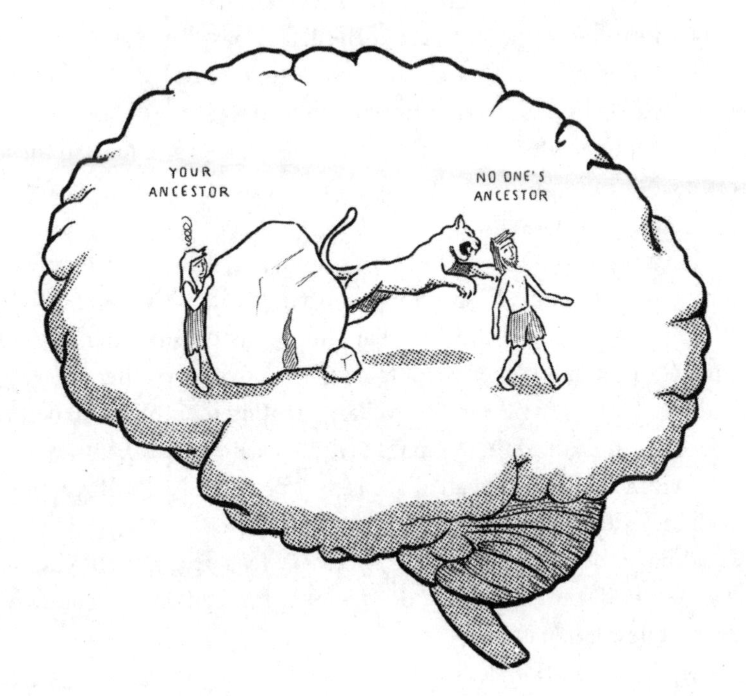

NERD NITE DC

DON'T TRUST YOUR BRAIN: Why Foreign Accents Are All in Your Head

by **Mari Sakai, PhD**

Have you ever wondered why foreign accents exist?

In most cases, humans across the Earth have the same anatomy for creating speech sounds. The lips close to make the sounds [p], [b], and [m]. The tongue tip raises behind the upper front teeth to create [t], [d], and [l]. And the middle of our tongue lifts for [k] and [g]. We also have *vocal folds* in the *larynx* that create vibration, called *voicing*. Touch your throat now and say a long "ah-hhh." Do you feel the vibration?

If you haven't seen your vocal folds in action, I recommend *running* to your computer and searching "vocal fold stroboscopy video." It is honestly amazing. The speed that those little guys flap, hundreds of times per second, will determine your pitch. Thicker vocal folds move slower, creating a lower pitch. Little-kid vocal folds are tiny and move fast, creating their high-pitched voices. Some speech sounds need that vibration, and are called *voiced*. Others only use air and no vibration, and are called *voiceless*. Voiced and voiceless sound pairs in English include [z] and [s], [d] and [t], [b] and [p].

If we all have the same body parts necessary for speech, then why can't we all make the same sounds? Where does a foreign accent come from when we try to learn a new language?

To answer that question, let's start with the first language.

By early childhood, we have established a robust set of all the sounds, or *phonemes*, we need to speak our native language. Within a few years, we've heard these phonemes billions upon billions of times. We implicitly know how to articulate each phoneme, how often it appears in the language, the words in which it appears, how the sound changes with different accents, and which sounds are its closest acoustic neighbors. These categories are so strong and so established in our brains that they essentially become a lens through which we hear all speechlike sounds. We are so efficient at processing speech sounds that we can hear in imperfect conditions like loud restaurants, we can fill in the gaps when certain expected sounds might be missing, and we can dump extra acoustic information that our phoneme guide tells us is not important. We are expert listeners of the phonemes in our first language.

Zoom out a little bit, and we come across "sound rules" that we follow in our first language without even realizing it.

If you're near another person, ask them to say these two sentences aloud:

> The tree drawing on my desktop is four months old.
>
> This spam number texts me every day.

In the best-case scenario, your friend will not be suspicious and their speech will be pretty natural. This mini-experiment might not work on someone (like you, the reader) who has been primed to be overly aware of their own speech.

How did your friend say the word *months*? Did you hear a [th] sound? What about the word *desktop*? Did you hear the [k]? Most English speakers, in casual, regular speech, will not pronounce the [th] or [k] sounds in these words. Why? English does not like three or more consonant *sounds* (not letters) in a row. So the middle ones have to go! "Desktop" becomes [destop] and "months" becomes [mons].

C C C → C ~~C~~ C

D E S K T O P → D E S ~~K~~ T O P → D E S T O P

M O N T H S → M O N ~~T H~~ S → M O N S

Did you notice the other long consonant string in the second sentence? Hint: It's "texts me." Second hint: The letter *x* is really two sounds, [k] and [s].

T E K S T S M E → T E K S ~~T S~~ M E → T E K S M E

It's just easier to say things that way, right? And everyone still understands exactly what we're saying.

These sound changes that occur in regular patterns are called *phonotactic constraints*. Every language has them, and it just means that the language has some preferences for how to put the sounds together. English regularly uses the sounds [b] and [n], and it is physically possible to say them together (think of the word *hobnob*). But American English speakers would never put those two sounds together at the beginning of a word. "Bnob"? Never! I can't even say it without adding a little vowel in between. That's a phonotactic constraint.

Remember, not only do we have an established and very strong phoneme inventory for our first language, we also have an established and very strong phonotactic constraint system.

Now try to learn a new language after decades of that stronghold in your brain. New phonemes from the second language are going to want to join the party. They'll need to shove in, pile on, and rip some of your existing phoneme categories apart. Guess what? New phonotactic constraints want to join, too. But you'll need to go ahead and suppress your first-language phonotactic constraints when you didn't even know you were following them in the first place.

A good example of first- and second-language phonemic inventories clashing involves vowels. American English has 11 vowel sounds, but let's just focus on two: the vowels in *sheep* and *ship*. In that same acoustic space, Spanish only has one vowel that's similar to the English [ee] sound. That's why many Spanish speakers of English might say "cheap" instead of "chip" and "sheet" instead of "shit." In order to say that "ship" vowel, they need to establish a new vowel category in that space. And it's hard. It takes thousands of times hearing that sound in order to establish a new category. And even when it begins to be established, it might be a weak category for a very long time.

If you're an American English speaker trying to learn Korean, you might be surprised to learn that where there was once just one [s] sound, there will now be two. I have tried so many times to hear the difference between the two Korean [s] sounds with my Korean friends, and I have never been successful. There are definitely acoustic differences, but my American English brain doesn't want to pay attention to them. For my brain, the two Korean [s] sounds are both good versions of my one English [s] category, and it won't hear of anything else!

Now let's talk about phonotactic constraints. Have you ever heard a Japanese speaker say "ice cream"? One phonotactic constraint of Japanese is that the language prefers consonant-plus-vowel combinations. "Ice cream" has the [s] consonant touching the [k] consonant, which is touching the [r] consonant. Japanese doesn't like that. So, it adds some vowels in there to feel more comfortable. Now it's [ai-su-ku-ree-mu].

Korean has a phonotactic constraint involving word endings. The language does not like to end words with *fricatives* and *affricates* like [ch] and [sh]. The fix is adding an extra vowel sound at the end. The English word *fresh* pronounced by a Korean speaker of English might sound like "freshi." "Watch" becomes "watchi."

There are certainly many other factors that complicate the story of foreign accents, but if you know these two concepts—phoneme inventories and phonotactic constraints—you're off to a good start. Are you a budding phonologist now? If so, go celebrate—with some *cheap ice cream*!

Mari Sakai has a PhD in Applied Linguistics. She is a faculty member at Georgetown University Law Center and a speech consultant focusing on improving communication in global and diverse settings. To learn more English pronunciation tidbits, follow her on TikTok @speechbites.

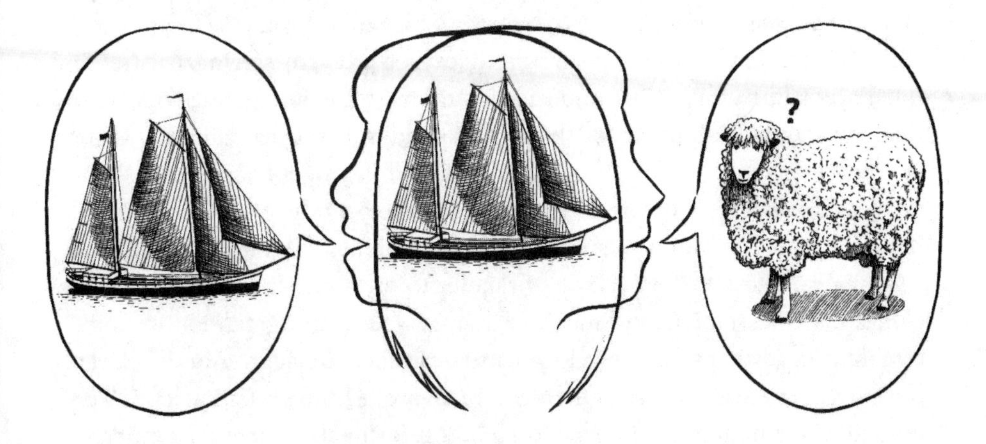

NERD NITE WINONA

LESSONS FROM *THE OREGON TRAIL*

by **Elizabeth Russell, PhD**

"You have died of dysentery."

Anyone who grew up in the 1980s or '90s knows the horror of those glowing, green words on a black screen. I spent day after day in my elementary school computer lab trying to figure out how to navigate *The Oregon Trail* without meeting such a fate. In doing so, I learned problem-solving, frustration tolerance, and history. Most important—and trust me on this one—I learned that you should always pay for the ferry instead of trying to caulk the wagon and float across.

I was a gamer long before I became a psychologist. My love of *Super Mario* has always allowed me an escape from the perils of life, childhood and graduate school alike. However, psychology has a lot to say about the impact of video games on development, and much of it is . . . well . . . not great. For example, research suggests that violent video games can increase aggression in players during game sessions. Though there are many legitimate and scientifically backed concerns about the impacts of gaming, fortunately there's also research indicating that video games have the potential to do good.

One positive aspect of video games is that they can teach resilience in the face of failure. In his studies of human development, psychologist Lev Vygotsky discovered the importance of providing challenges within the zone of proximal development, in which a task is not so easy that it is boring, but not so challenging that it causes someone to become frustrated and give up. A well-designed video game begins at a low, manageable level of challenge that teaches players the basics of the game. Then they can use those skills to move up the difficulty ladder, thereby developing more skills and preparing them for advanced tasks. In most modern games, if a challenge is just too much, players can typically run around, defeat some enemies, "level up," and then return to the task when they are stronger. The ultimate result of this is that players build up better frustration tolerance and persistence. Even when they struggle, they can keep working at it, and eventually they know they will be rewarded with the satisfaction of finally achieving their goal.

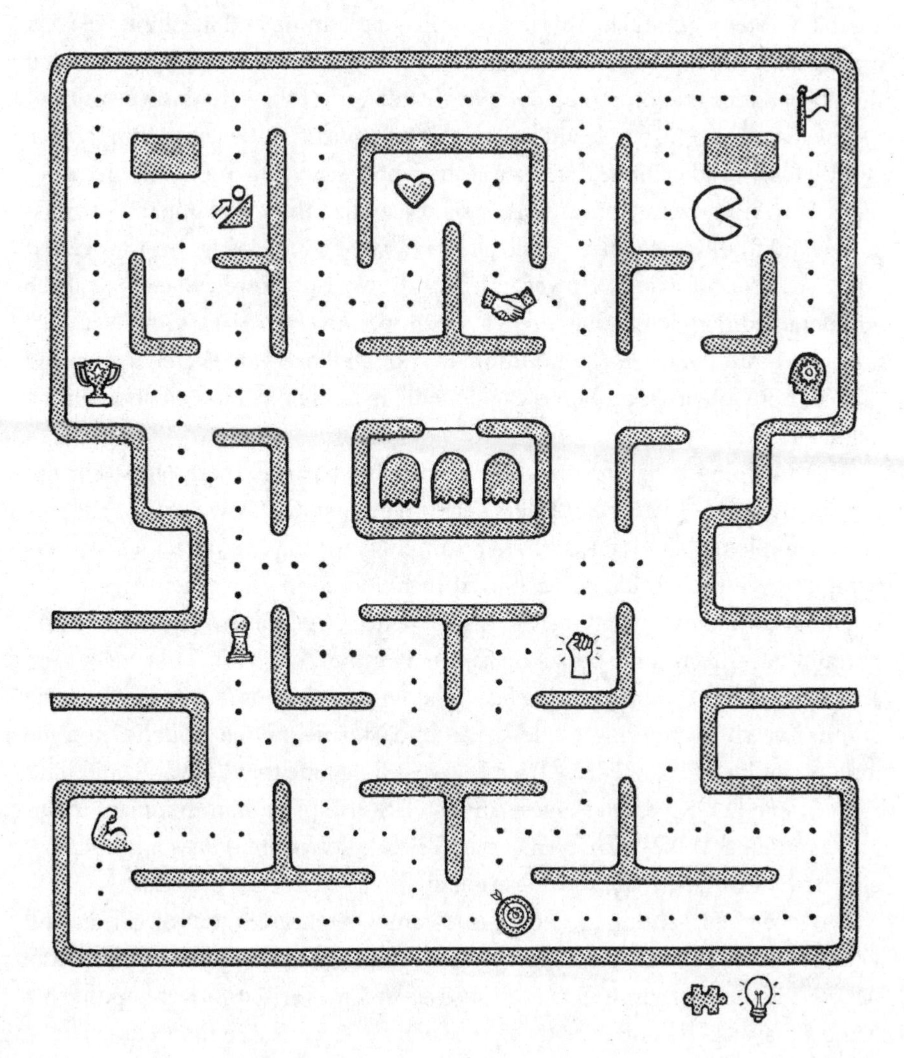

Further, video games help us with creative problem-solving. Often, there are multiple means by which a challenge can be defeated. Take, for instance, the modern PlayStation game *Horizon Forbidden West*, in which the protagonist, Aloy, fights to save a post-apocalyptic world overrun by aggressive machines. She has her standard bow, and the game can be won using just that, but Aloy has so many other tools at her disposal that it's a shame not to try them all. Stealth skills. Explosive traps and trip wires. Elemental arrows. She can even shoot heavy weaponry off machines and then use it against them. This type of choose-your-own-adventure strategy—common to many games—not only makes things more fun and inventive but also allows multiple options for players to try if they get stuck. This builds skills in planning and flexibility. Players can identify an approach that might work, take the steps to carry it out, and then tweak it or try something completely different if they are unsuccessful.

So what about the whole video-games-and-aggressive-behavior thing? Well, yes, that is a valid critique, but it's not the whole story. Dr. Tobias Greitemeyer of the University of Innsbruck in Austria researched the social impacts of video games and found that violent video games can be a source of aggression-related emotions and behaviors. However, his findings also showed that this can be counteracted if violent games are played cooperatively or if altruistic elements are added into those games. Additionally, not all video games are violent, and playing nonviolent video games can actually increase compassion and helping behaviors.

These helping behaviors can take place within games themselves. For example, in *World of Warcraft* players can join guilds that allow them to cooperate to complete difficult tasks, share resources, and simply connect with others from all over the world. When played in moderation, video games have the potential to facilitate relationships and expand social networks. I don't think I would have been quite so successful in *World of Warcraft* if not for a kind soul named IEatKittensForBreakfast, who stumbled upon an underequipped gnome "newb" wandering aimlessly around Stormwind City. She invited me into her guild, and now I have friends across the globe that I never would have met without her. Research suggests these relationships are commonplace in online gaming, and they can even serve to break down prejudices and increase empathy for others. Also, she doesn't actually eat kittens for breakfast.

The effects of gaming are much like games themselves, full of heroes and villains and a lot of gray area in between. While there are plenty of potential negative impacts of gaming, many of those outcomes are contingent upon what game a player plays and how they approach it. Given the right circumstances, gaming has plenty to offer beyond simple entertainment. In my own journey, I developed a lot of logic, self-control, and good old-fashioned grit, in addition to some serious emotion regulation. I have rarely been so proud as I was when I beat the dysentery and reached the end of the Oregon Trail. Now if you'll excuse me, I have a date with *Donkey Kong*.

Liz Russell is an associate professor of psychology at Winona State University. She holds a PhD in Counseling Psychology, and she's beaten Horizon Zero Dawn *on "ultra hard" four times.*

NERD NITE ST. CLOUD

SYNESTHESIA: Hearing Colors and Tasting Sounds

by **Rebecca Woods, PhD**

"You've got synesthesia? Oh my God! Is it contagious?"

If you've never heard of synesthesia, don't feel bad. You're not alone. Synesthesia isn't a disease, nor is it something that should be followed by "Gesundheit!" Synesthesia is a condition in which a sensory perception that typically occurs in one sense occurs when stimulated by a different sense, such as seeing colors when sounds are heard. The word comes from the Greek *syn* or "union," and *aesthesis* or "sensation." Literally, it means union of the senses.

Although there is evidence of synesthesia throughout history, serious research on the phenomenon didn't begin until the 1970s. Researchers estimate that approximately one in 23,000 people have synesthesia. However, when a very liberal definition is applied, one that is expanded to include emergence in childhood and very subtle forms of synesthesia, the estimate is closer to one in four.

But let's stick with the more conservative definition. Synesthesia is not the same as a simple association. It's not "this makes me think of that." Rather it's "this makes me experience that." Hallmark criteria of synesthesia are that it's automatic and consistent. When tested and retested, synesthetes always experience the same thing (for instance, when I hear this high pitch, I always see a small white light, but when I hear this low pitch, I always see this large orange light). Synesthesia can occur between any two—or sometimes three—senses (such as sight and hearing, taste and touch) and can even occur within a sense (say, color-shape). Synesthesia can also occur with more complex stimuli. Someone with color-emotion synesthesia will see colors when they detect emotions in others based on facial expressions and body cues. And you thought those people who said they can see auras were making it up. Think again. So far, there are as many as 80 documented forms.

So, wondering who has synesthesia? Just about anyone can be a synesthete. Men, women. Lefties, righties. However, there may be a genetic predisposition. If one family member is a synesthete, it's likely that others in the family have synesthesia, too, although their experiences likely differ. There are a lot of famous synesthetes, such as Marilyn Monroe and Nikola Tesla. And although there are many artistic synesthetes, including Wassily Kandinsky and Billie Eilish, the rates of synesthetes in the arts is about the same as rates in other careers.

While it may seem cool to have synesthesia, some synesthetes find it debilitating. One of the most common forms of synesthesia is a within-sense form called color-grapheme synesthesia. Children who have this or similar types of synesthesia often have trouble in school because the colors they are seeing don't match those that others are seeing. Remember kindergarten when you had to say what color the triangle was? A color-shape synesthete may not see the color the rest of the class does. And to further complicate things, many synesthetes have no idea that others don't experience what they are experiencing; they don't know their experience is different. Imagine the frustration! But it's not all bad. Many synesthetes have an advantage over others because, for example, their dual sensory experience makes remembering things easier or finding items faster. Color-pitch synesthetes may have an easier time singing because they can see the distinction between pitches or can literally see when they hit the right sound. One of the classic experiments for color-grapheme synesthesia is having participants look at a field of two letters, such as *T*'s with a few *L*'s thrown in. Since these two letters are experienced as different colors in color-grapheme synesthetes, the *L*'s will immediately pop out from the *T*'s.

So what's going on? Is this magic? Of course not. Our perceptual experiences occur as a result of our highly complex brains, and because sometimes neural processes can go a little off kilter. Neural imaging shows that in synesthetes' brains, activation occurs in the part of the brain usually specialized for processing the stimulus, but activation also occurs in the areas specialized to process the paired sense. For example, if an auditory-visual synesthete hears a sound, the auditory cortex "lights up," but so do areas of the visual cortex. Both areas of the brain are processing a stimulus that would typically be processed by only one of the areas. According to the cross-activation hypothesis, this occurs due to cross-wiring of neural connections in a way that causes perception to occur for both senses. This hypothesis is supported by research suggesting that the senses might be less distinct in the prenatal and infancy periods and that they differentiate in the early months of life. In other words, all babies are synesthetes, but as the brain continues to develop, those connections get pruned away in most people. Another explanation is the disinhibited feedback hypothesis in which inhibition processes in the brain that prevent activation to the "wrong" stimulus (top-down processing) are not working as they typically do. In this case, the idea is that the cross-wiring is present in everyone, but what isn't working properly is the inhibition process. Other explanations have been proposed and, although it's not clear which of these explanations is correct, research is ongoing and continues to enlighten us about this phenomenon.

If you think you might be a synesthete, you can contact one of several labs around the world where synesthesia is studied. If you're not a synesthete but you want to know what it's like, you may be able to get something close to it. Synesthesia can't exactly be induced, but synesthesia-like experiences can occur as a result of training (repeated associations) or hypnosis. Some people have reported synesthesia-like experiences as a result of chemical exposure (such as psychedelic drugs), sensory deprivation (a damaged optic nerve or missing limbs), and traumatic brain injury. But for the safest experience—you know, one that doesn't rely on a brick to the head—look for demonstrations of synesthesia on the web. There are a lot of fun ones out there!

Rebecca, like most nerds, loves learning about pretty much every-thing! *She's an artistic scientist who was once a professor of babyology and now works remotely for a company that studies rare diseases. Her favorite nerd-out topics are those that pit perception against "reality"—and* Star Wars.

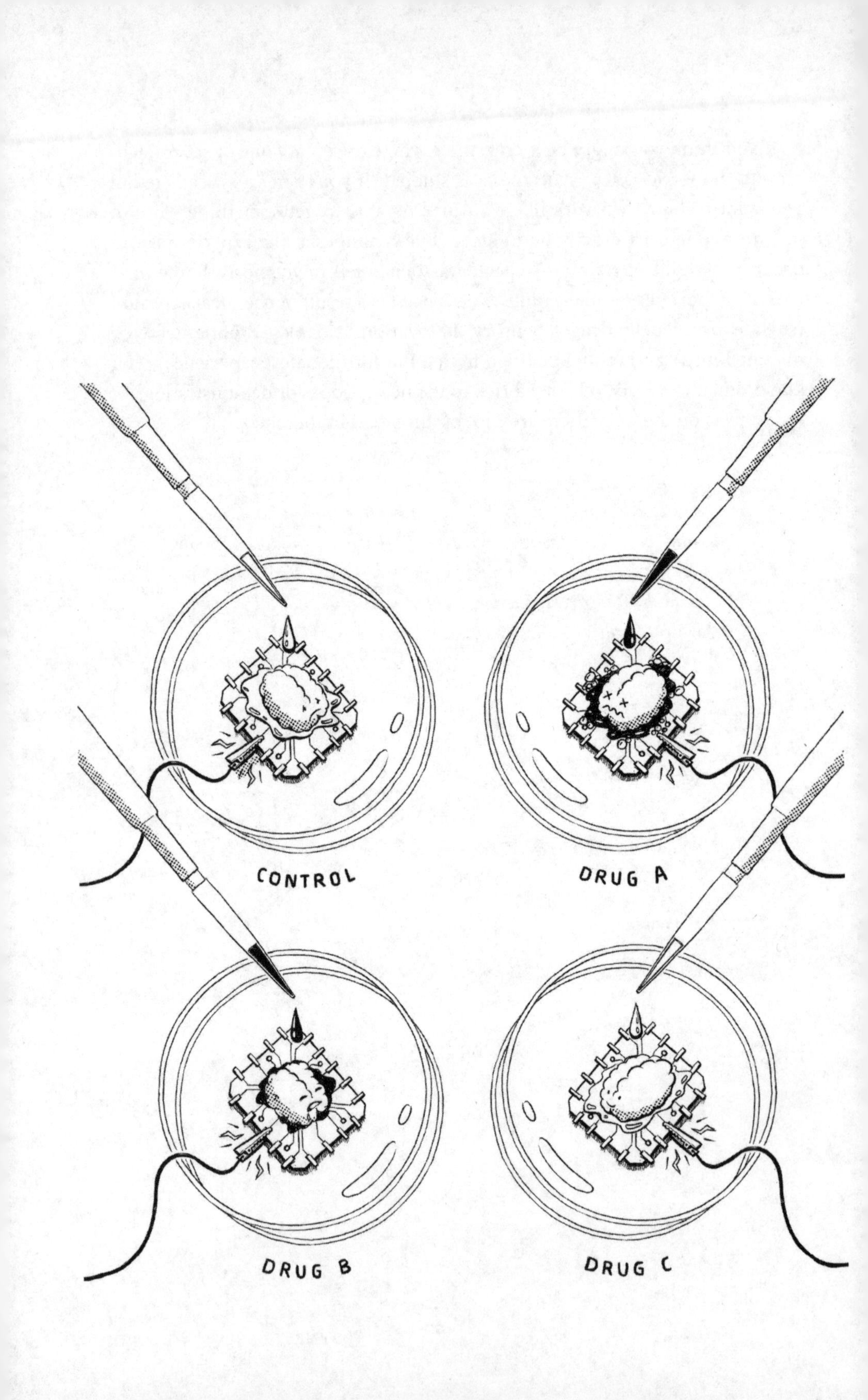
CONTROL
DRUG A
DRUG B
DRUG C

NERD NITE ORLANDO

BRAIN ON A CHIP: The Ethics of Brain Experimentation

by **Max Jackson**

How do the skulls of ancient civilizations drive us to develop technologies that will help the brains of those in upcoming generations?

Neuroscience today is crawling its way out of its infancy and is sort of standing unsteadily in its cranky toddler period. It's been thousands of years in the crib! Neuroscience, in some form or another, has been around for as long as there have been humans. About 5 to 10 percent of our earliest ancestors' skulls feature giant holes missing from them, indicating that some of the oldest humans took a shot at brain surgery. In fact, it's the oldest form of surgery that we know about.

And even though the thought of brain surgery conducted thousands of years ago may sound rather dicey, the good news is that many of these skulls show signs of healing and recuperation, indicating that many of our earliest ancestors survived our extremely literal first stab at neuroscience. But the bad news is that we really haven't done a whole lot better than that over the past few millennia. The tools just really weren't there. For example, we didn't get the first useful modern brain sensors until very early in the 20th century, and even those didn't really come into their own until the advent of digital computing about half a century later. But now we're really on fire, technology-wise, as we're developing brain sensors and brain stimulators and figuring out increasingly powerful ways to analyze the mountains of data we're collecting.

But neuroscience still has one central problem: You only have one brain! If we accidentally break your brain, then we break you, and you'll either lose your mind or die. That's real bad news! What's more, your brain has its own agenda—it tries to correct itself whenever we do something to it. So how then do we study and repair the most vital of the organs that has a mind of its own? It's pretty simple actually—just build another brain.

So . . . how do you build another brain?

The same way you build anything else, pretty much: You start with the smallest parts and then build up from there as best you can. Now, in the case of the brain, the smallest part that we (kind of) have our grasp on is a specialized cell called a *neuron*. It's a cell that's specialized in sending signals and connecting with other neurons. You have about eighty to a hundred billion neurons, all of which communicate across connections called *synapses*, and you have about 100 billion of these in your brain. For comparison, your brain has more connections than there are stars in the Milky Way galaxy. That's a lot of complexity! The good news for us neuroscientists is that neurons do the same things whether they're inside the brain or outside—which means we can grow neurons outside a brain and they'll still send signals and connect with one another.

We do this through a special practice called *in vitro neuroscience*. The term *in vitro* is Latin for "in glass," referring to the flasks and test tubes and petri dishes that you'd find in a science laboratory. This practice gives us an opportunity to study neurons on a level that we absolutely cannot reach whenever they're in a person's brain doing vitally important things for them.

It's certainly not good if a scientist tries to insert a sensor into a normal brain, because the body treats it like a giant splinter and tries to destroy it. Therefore, we've devised a way in which to grow neurons right on brain sensors themselves. We can grow neurons on little devices called *micro electrode arrays*, which are made of metal that enables them to sense electrical signals and even deliver electrical signals back—we can grow a brain on a chip! This gives us an awful lot of advantages whenever we're doing neuroscience. Like I said: If we break your brain, we break you. But if we break a brain-on-a-chip: Oh dang, I guess I'll build another one, or maybe another 12.

For example, if you get sick, we can either test one treatment at a time on you and hope it doesn't drive you insane or kill you, or we can test 12 drugs at a time on these little brains-on-a-chip with no real consequences. And as we further personalize medicine, we'll eventually be able to make brains-on-a-chip out of not just any neurons, but out of *your* neurons.

As of now, animal testing has gotten us pretty far, but research animals are different enough from humans that animal testing results don't always necessarily generalize to human biological systems. Research animals are similar enough to us to feel pain and do other things that force us to kind of walk an ethical tightrope whenever we work with them, but if we're able to work with *your* cells with *your* genes and *your* history and *your* weird new problems, then we'll be able to discover treatments that are personalized for *you*.

What really makes this exciting for me is that it helps us deal with one of the great and terrible constants of human life: *decay*. This is where the real scariness is for me: When your body falls apart your brain falls apart, and when your brain falls apart your mind falls apart, too. So with your mind go your memories, your wit, your relationships—a lifetime of experience is on the line here. But there's hope, for me, in what I shared since in vitro neuroscience opens the door to our being able to build replacement parts for your brain out of neurons with your genetic material.

We're working to take things from *personalized* medicine to *regenerative* medicine, and my hope is that this technology will mature as we mature, thereby ensuring that all of us alive today can pack our life with as much experience and joy and love as we possibly can. My hope for this is that as this technology develops it'll enable us to give our grandchildren's grandchildren a life that is better than those of us alive today could possibly imagine.

Max is a data engineer at InformData. He is passionate about technological and personal growth and is required by intergalactic TED® law to mention that he has a TED® Talk about neuroscience.

NERD NITE LONDON

HOW WE BECOME DISGUSTING (SOME MORE THAN OTHERS)

by **Dr. Richard Firth-Godbehere**

Something which has never occurred since time immemorial: A young woman did not fart in her husband's lap.

—Sumeria, ~1900 BC

Disgust is an emotion shared by everyone. Or, at least, that's what Paul Ekman wants you to think. Professor Ekman traveled all the way to Papua New Guinea in the late 1960s to see if the Fore people—also known for kuru, a rare neurogenerative disorder spread by eating human brains—experienced the same set of six feelings as everyone else he'd tested. He did that by showing them pictures of faces and asking them questions like, "What face would you pull if you find a rat in your doorway?" It was the same test he'd used on people from all over the globe, but this time he had a group of humans who were so cut off from the world, they hadn't even seen *I Love Lucy*. The Fore, it appeared, did feel the same six emotions as everyone else—anger, fear, happiness, sorrow, surprise, and disgust.[1]

Unfortunately, there are a few issues with the experiment, not least of which is that the Fore weren't untouched people by the West—the people who discovered kuru had touched them 20 years earlier. Also, it turns out that disgust is understood differently by different cultures. For example, when was the last

1 Paul Ekman, "Universal Facial Expressions of Emotion," *California Mental Health Research Digest* 8, no. 4 (1970): 151–58.

time you ate a human brain? The English-speaking world tends to think of disgust as feeling nauseated and yucky. The Germans think of disgust, or *Ekel*, as a need to move or get away from something. The French think of *le dégoût* as things of which we have too much. Even beautiful works of art become garish when you have too many in one space. But underneath the cultural differences, there is indeed a basic feeling of revulsion that all humans share.

One theory as to why came from Disgustologist Valerie Curtis. She posited that we feel revulsion whenever we see, smell, taste, touch, or hear something that might contain pathogens or parasites. Evolving a strong feeling for rotting things and things crawling with mold or bacteria has obvious evolutionary advantages. The biggest being you don't die, and you get to have offspring. Professor Curtis called this pathogen avoidance theory, or PAT.[2]

How disgust works in the brain isn't completely understood, but we have some clues. New ideas appear all the time, from gut hormones to a suppressed amygdala. One lead is the role of a nonapeptide known as oxytocin. You might have heard of oxytocin as "the love chemical." Really, oxytocin increases whenever we are interacting with or thinking about anyone we care about, whether a romantic love, a friend, or a member of our family. But when something or someone disgusts us, oxytocin is repressed. The brain decides this thing or person is to be pushed away or avoided.[3] Another important part of the disgusting brain is the *insula*. Generally, this bit governs how strongly we feel something—its valence—and is particularly active when we are being revolted. Most interestingly, it also lights up in fMRI scanners when people are presented with morally reprehensible acts, so there seems to be a physical link deep in our gray matter between morals and most people's reaction to eating fresh brains.[4]

As a social group, we've come to develop agreed-on healthy patterns of expected behavior that are beneficial for the group as a whole—we tend to call them morals. When someone in the group starts acting weird, or in a way that might damage the group, they cause the same feelings of revulsion. We become disgusted by their actions. Experiencing these things makes us feel like we need to clean ourselves, like we've been infected. Even the thought of wearing, say,

---

2 Valerie Curtis et al., "Disgust as an Adaptive System for Disease Avoidance Behaviour," *Philosophical Transactions of the Royal Society: Biological Sciences* (February 12, 2011).

3 M. Kavaliers, K.-P. Ossenkopp, and E. Choleris, "Social Neuroscience of Disgust," *Genes, Brain and Behavior* (2019).

4 P. Wright et al., "Disgust and the Insula: fMRI Responses to Pictures of Mutilation and Contamination," *NeuroReport* 15, no. 15 (October 25, 2004): 2347–51.

Adolf Hitler's sweater causes what's known as a *contagion heuristic*.[5] Somehow, we are so sensitive to revolting things that we think we can be made impure simply by touching something an evil person has touched.

Another way this moral-physical link has been demonstrated is through the Disgust Scale-R Test. Developed by psychologists Jonathan Haidt, Clark McCauley, and Paul Rozin, it asks questions like, "I might be willing to eat monkey meat, under some circumstances" and "You see a man with his intestines exposed after an accident."[6] Each question is answered on a scale, and those who are more easily revolted by the questions tend to be politically extreme. It looks like anybody who is political to the point of being tribal sees those with opposing, or even less extreme, views as something impure, to be cleaned before they cause infection.

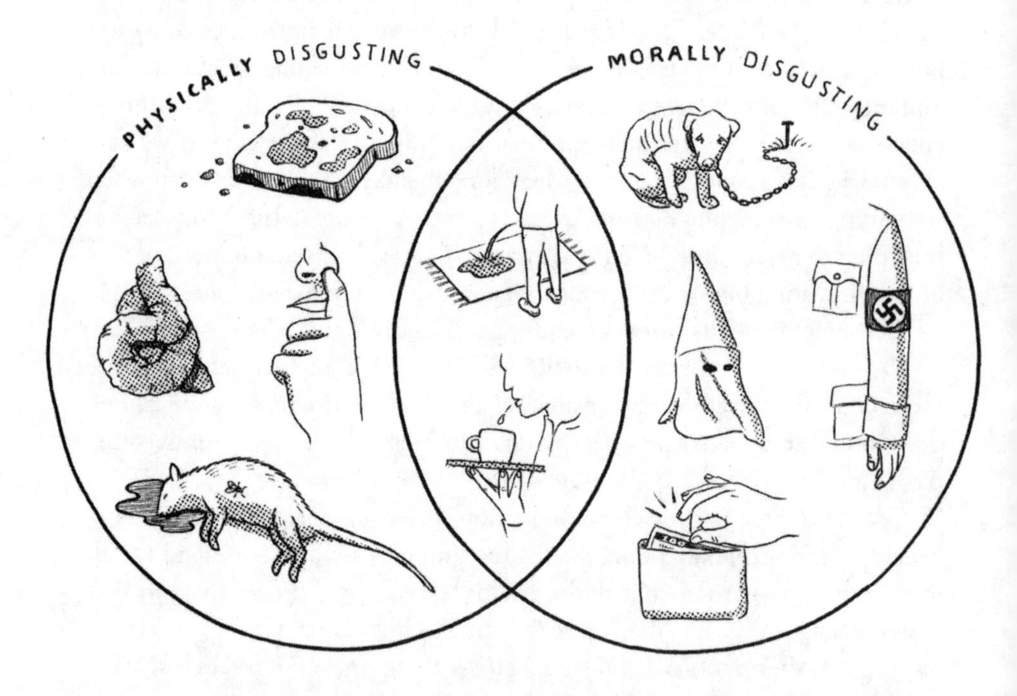

Because of this, disgust can be dangerous. Famously, the Nazis made films like *The Eternal Jew*, comparing the Jewish people to rats and dehumanizing

5 Carol Nemeroff and Paul Rozin, "The Contagion Concept in Adult Thinking in the United States: Transmission of Germs and of Interpersonal Influence," *Ethos* 22, no. 2 (1994): 158–86.

6 https://pages.stern.nyu.edu/~jhaidt/disgustscale.html.

them so it would be easy for the citizens of the Third Reich to accept the Final Solution. Almost every act of genocide involves someone making someone else out to be revolting and inhuman. Even as far back as the witch trials of the sixteenth century, women—and it was mostly women—were depicted by artists like Albrecht Dürer as "abominations." And yes, unfortunately judging women by their looks is not a new thing. Witches were represented as older and less firm, with those bits of them sagging that will sag on all of us one day, and usually doing something morally wrong. Ninety percent of the time that meant either a forbidden sex act, riding backward on something, or eating babies. *Abomination* was thought of as a sort of religious revulsion—the feeling you had when you came across a sinner. Witches were, at the time, the worst possible kind of sinner.

Using disgust as a weapon to dehumanize has a long and awful history. But so, too, does finding disgust deeply fascinating, even funny. The oldest known joke is a roughly four-thousand-year-old Sumerian fart joke. No one knows quite why disgust mesmerizes and titillates us so, but I'm hoping it's what got you to the end of this chapter.

Richard Firth-Godbehere, PhD, is one of the world's leading experts on disgust and emotions. He is Distinguished Professor of Liberal Arts and Humanities, Woxsen University, India, and an honorary research fellow at the Centre for the History of the Emotions, Queen Mary University of London. His award-winning interdisciplinary research walks the line between history, psychology, linguistics, philosophy, and futurism. He examines how understandings of emotions change over time and space, and how these changes can influence the wider world.

Bodily Fluids

What do you say to someone having difficulty peeing? Urine trouble.

What do you call an accountant taking a pee? A math whiz.

What kind of pirate pees on you? Yaaaaaaaaaaaaarrrrrrr Kelly.

Mucus puns? Don't even goo there.

What's brown and sits in a courtroom? Jury doody.

Why did the condom fly across the room? It got pissed off.

I tried to help after the spit massacre, but there were no salivas.

My spit is so spicy, I call it phlegm brûlée.

Donating blood is A-positive thing.

When you're walking in the street and you feel something neat . . . diarrhea, diarrhea.

When you're walking down the hall and you feel something fall . . . diarrhea, diarrhea.

When you're sliding into home and you feel something foam . . . diarrhea, diarrhea.

When you're solving a math equation and you feel a wet sensation . . . diarrhea, diarrhea.

When you're digging in the dirt and you feel something squirt . . . diarrhea, diarrhea.

Grow up already!

Bladder control. Poop in space. Antibacterial soap and super bacteria. Doody from our toilets running into our streets. Milk from, well, pretty much every living thing. Yes, bodily fluids can be gross and make us squeamish, but heck, we all have them. Yes, even *you*! Phlegm, mucus, blood, urine, poo (when liquidy), pus, semen (well, about half of you), and so much more icky, drippy runniness is happily flowing through all of us. Right now. This chapter dives into the fluids that don't just make us squirm but make us human. Well, more like ewww-man. Zing!

—Matt

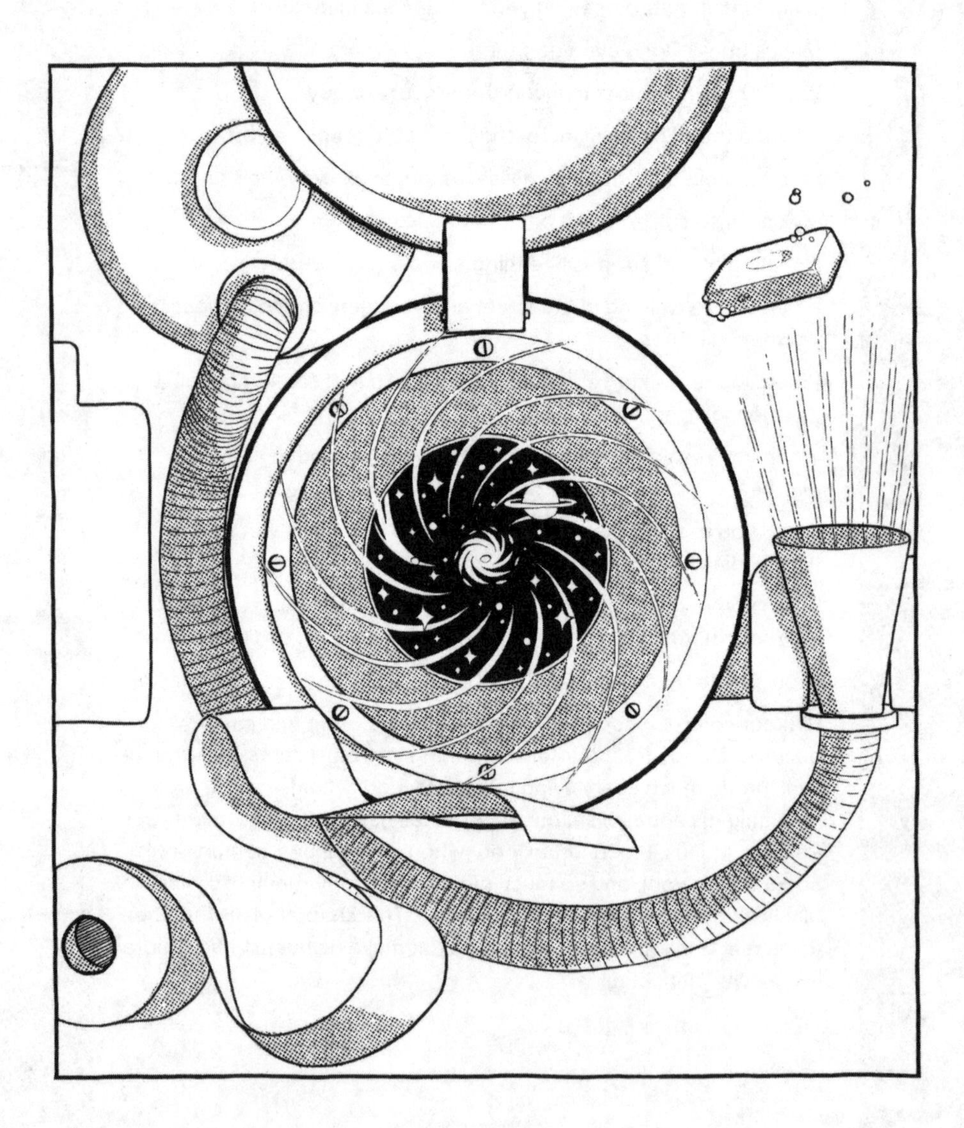

NERD NITE ORLANDO

TO BOLDLY GO: Dealing with Poop and Pee in Space

by **Brendan Byrne**

The *Apollo 10* mission in May 1969 was a critical dress rehearsal ahead of NASA's first lunar landing. Six days into the mission, as the crew of three astronauts were making the long journey to the moon, they encountered an unidentified flying object—*inside* their capsule.

"Give me a napkin quick," said commander Tom Stafford. "There's a turd floating through the air."

It wasn't the last sighting of an unidentified lunar log. A few minutes later, crewmates John Young and Gene Cernan spotted another and argued over who it belonged to. "I don't know whose [sic] that is," said Young. "I can neither claim it nor disclaim it."

It was quite easy to lose track of waste during NASA's early days of space exploration. Urine was collected using a diaper, a condom-like attachment, or a bladder-like collection device worn under the astronauts' clothes. They were routinely emptied, and the liquid was released outside the spacecraft—creating a glimmering cloud of frozen astronaut piss.

Solids, on the other hand, were kept on board in bags. The astronauts would align their "waste ports" to a hole in a plastic bag and perform the evacuation. The bag was sealed, and the astronauts would knead into the feces a special disinfectant to help with the spread of germs and smells. Because of the weightlessness in the capsule, and the occasional docking issue, sometimes some waste escaped.

While gross, this worked for the shorter mission durations of *Apollo*. But as NASA looked toward longer-duration flights, another method would be needed. Toilets became the standard method for waste management when people began spending more and more time in space. Skylab, the United States' first space station, was equipped with a toilet.

Using a cosmic commode isn't as easy as using one here on Earth. For one thing, our earthly poos have the advantage of gravity—the waste falls down to where it needs to go. In space, astronauts don't have that luxury. Instead, space toilets utilize negative pressure to pull the waste away from the astronaut's bum and into a can for disposal. This vacuum also helps with smells.

But the space toilet isn't just for collecting waste for disposal. It also collects waste for reuse. That's right. *Reuse.*

When working and living in space, you must take everything with you. Clothes, food, fuel, and water. Transporting all those resources doesn't come cheap, so what if you could reuse some of those things?

When building the International Space Station, engineers did just that. Starting in 2009, astronauts on the ISS began recycling their urine, reclaiming about 75 percent of the water from their own pee. Since then, upgrades to the system have made it more efficient. And despite the source, astronauts don't seem to mind. In fact, many have said it's the best-tasting water they've ever had.

Being able to recycle these resources will help us explore farther into our solar system. NASA is developing its *Orion* space capsule, the first spacecraft to take humans into deep space since the *Apollo* program. A new capsule is going to need a toilet, and work on that technology has already begun. The agency developed a new toilet—it's smaller, more efficient, and built with 3-D–printed parts that can easily be manufactured in space should an astronaut plumber need to fix a clogged commode.

A version of that new toilet was sent to the International Space Station in 2020. NASA instructed astronauts to stress-test the systems and "give it all they got" with plans to install a similar toilet on the *Orion* space capsule, for use by astronauts heading to the moon.

Another advantage of NASA's new toilet: It's better suited for female astronauts. For decades, the NASA toilet wasn't optimized for female anatomy. This new moon-bound toilet has hardware that allows NASA's women astronauts to use the toilet much more easily than previous designs did, including what's known as a dual op: doing a #1 and #2 at the same time (space travel is all about efficiency, right?).

Waste management doesn't stop at the toilet. Other scientists and engineers are working on ways to reuse more waste—including feces—for fuel and even water. NASA's Trash to Gas technology would take that waste and incinerate it, creating valuable water that could also be broken down into hydrogen and oxygen, two key sources for rocket fuel. This will be crucial as humans explore the moon and even Mars. One person's trash is a spaceship's gas.

What started as a floating mystery back on that *Apollo 10* mission is now turning into a valuable resource that will fuel our exploration of the solar system. And what we learn from these experiences can be used down here on Earth. As our planet becomes overwhelmed with trash, this technology could be used to recycle that crap and provide valuable resources to all the crew here on spaceship Earth.

Thanks to a focus on waste in space, the space toilet will boldly go where no toilet has gone before—to the moon, Mars, and beyond.

Brendan Byrne is a journalist at WMFE in Orlando, Florida, and NPR contributor, reporting on space exploration. He hosts the weekly podcast and radio show Are We There Yet?, *which explores space exploration.*

NERD NITE GREENVILLE, NC

MILK! YOU'LL SEE IT EVERYWHERE ONCE YOU KNOW WHAT TO LOOK FOR

by **Kristen Orr**

Back in 2019 when I was starting my journey to study a form of milk found in pigeons and doves, I could not have anticipated the truly weird, years-long rabbit hole I was about to go down. It took me way beyond the world of pigeon crop milk (a semi-solid substance that I'd argue more closely resembles saffron rice than cow's milk) and forced me to question the very definition of the word *milk*. We all *think* we know what it is, right? We think of it as a white liquid that comes from mammals, or maybe from almonds/oats/soy/rice/coconuts/any plant under the sun (but *damn* does this get people in the dairy industry heated). That's only scratching the surface, though. While I don't have strong feelings about what it says on cartons in the grocery store, I do feel as if we need to broaden our definition of the word as a biological term.

So if you've got an open mind and are down to be a little grossed out, please join me on a brief survey of milk in its many freaky and fantastical forms.

Let's first start with a definition. While the biological definition is still hotly debated (what a strange world we live in), it's generally agreed upon that milk is any nutritious substance that is produced in the body of a parent and fed to developing young. There are some caveats here: (1) it must represent a nutritional sacrifice by the parent, (2) it can't kill the parent, (3) it can't be collected and regurgitated, (4) it can't be yolk-based, (5) it must be ingested by the young instead of absorbed, and (6) it must be vital to the survival and well-being of the young.

Following this definition, milk is produced by:

Mammals: The (perhaps undeserving) VIPs, the ones that put this stuff on the map. The females of everything from whales to platypuses secrete liquid milk from somewhere on their ventral surfaces, which their babies suck up.

Birds: Male and female pigeons, flamingos, and male emperor penguins make a chunky stuff in a throat pouch called the crop. They then regurgitate it into the open mouths of their newly hatched chicks. Yum.

Amphibians: There's an animal called a caecilian that lives underground and looks like a worm. A female will grow a fatty outer layer of skin and then allow her babies to PEEL. IT. OFF. HER. And eat it. She then grows another layer in a few days like it's NBD and lets them do it again. There's also the Nimba toad and the alpine salamander, both of which feed their babies milk within the uterus and later give live birth to them.

Fish, sharks, and rays: Male and female discus fish secrete some extra mucus that is extra tasty for their offspring. It's secreted all over their flanks, and the newly hatched fish nip it off their parents. There's also a whole bunch of sharks and rays that make milk in their uteruses to feed to their unborn babies (great white shark moms, for example!).

Insects, arachnids, and crustaceans: Where to even begin with these guys . . . We've got several milk-producing flies like the tsetse fly, louse fly, and bat fly (don't look that last one up unless you enjoy horror). And there are several groups of bees and wasps, a species of earwig, the Pacific beetle cockroach, and several isopods, like the Antarctic giant isopod and the common woodlouse, all of which produce milk. We've also got the milk-producing pseudoscorpions and even a few real-deal scorpions. Most of these produce milk for their babes within uterine tubes or broodsacs, the exception being the bees and wasps. They feed newly hatched larvae a milk (often referred to as royal jelly) produced by glands in their heads and secreted through an opening near their mouths.

Everyone else: Since there are more milk-producing animals than there is room here, everyone else gets unceremoniously dumped into this last group. Here we have a few sea cucumbers and starfish, various mollusks like the violet sea snail and the swan mussel, and a whole bunch of worms and wormlike creatures (the cutest being the velvet worms). Most of them make milk within ovaries, oviducts, or brood pouches, then feed it to their babies to prepare them for the big wide world.

So why do we care? Why does knowing about milk matter? Two reasons: (1) it's gross and therefore cool and (2) it's a humbling reminder that we're not so unique. We like to think that, as mammals, milk production is this genius parental-care strategy that only we as a group have landed on. But as the research has shown, milk is everywhere—you'll see that the world is practically drenched in the stuff now that you know what to look for.

Kristen Orr is a freelance artist and designer based in Arlington, Virginia. She has designed exhibits for science museums, created scientific illustrations for books (including this one), and studied design and biology.

NERD NITE FARGO

TRICLOSAN: It's Not the Bacteria but the Soap That's Going to Kill You!

by Dr. Graeme R. A. Wyllie

When it comes to personal care products, throwing in the word *antibacterial* sure seems like a sales booster. Of course, when it comes to soap, the evidence for whether antibacterial products are any more effective than regular soap and water at killing bacteria on surfaces such as your hands has never been definitive. But from 2009 on, and coinciding with concerns over bird flu and swine flu (H1N1) (both viruses and unaffected by antibacterials), the market for antibacterial soaps and related products exploded. And what better material to use than one that had been known for at least forty years, was relatively cheap to produce, and had proven antibacterial properties. Since 5-Chloro-(2,4-dichlorophenoxy)phenol doesn't exactly roll off the tongue, this material is generally listed as either triclosan or Irgasan on ingredient labels of products such as antibacterial soap.

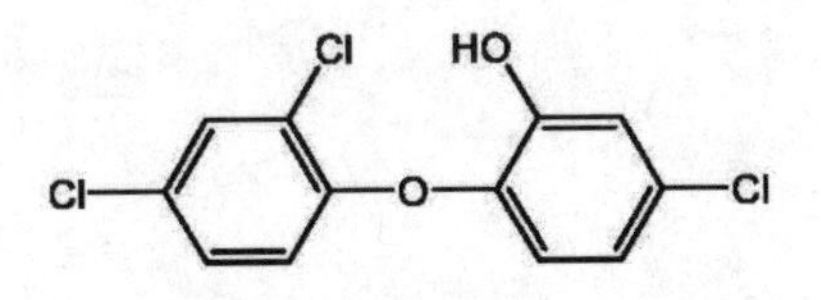

Chemical structure of 5-Chloro-(2,4-dichlorophenoxy)phenol

Triclosan itself was originally synthesized and patented in the mid- to late 1960s and by the early 1970s was popping up as the antibacterial component in mouthwash, toothpastes, shampoos, and soap, as well as being impregnated in materials like sutures and sterile surfaces such as cutting boards. In pure form, it's a white powder that has horrible solubility in water but does dissolve more readily in an amphiphilic (both hydrophobic and hydrophilic) material like a soap. The hydroxyl (-OH) group on the six-membered carbon ring is reminiscent of phenol aka carbolic acid, an early antiseptic dating back to Scottish surgeon Joseph Lister, widely considered to be the father of sterilization (not a reference to the name of one of his kids). The addition of the chlorine groups, however, ups the ante on the toxicity of the molecule, and it was found to have significant antibacterial properties that essentially prevent bacterial growth by putting the boot in their fatty acid synthesis pathways.

If all that sounds great, it was no surprise that by 2010, thousands of products were readily available containing this wonderful antibacterial additive, and in the case of some liquid soaps, the amount present ranged from 0.1 to 1 percent of the weight of the material. No worries there except that a wave of reports in scientific literature started cropping up saying maybe our miracle chemical isn't as great as we think.

It's only on our hands, though, right? Not the case, as it turns out triclosan can be absorbed through the skin as well as direct ingestion (washing your hands before you eat becomes scarier all of a sudden), and numerous studies reported detection of triclosan in our urine, blood, and even breast milk. Still nothing to worry about except for reports of endocrine disruption, muscle impairment, liver tumors, and even promotion of nasal colonization (great band name, by the way) by *Staphylococcus* in rats. Combining the potential issue of developing antibacterial resistance with the fact that triclosan did not discriminate between harmful and benign bacteria meant suddenly our antibacterial products didn't seem like such a great idea.

Of course, triclosan is also the gift that keeps on giving, and the massive quantities we were using didn't just disappear like your average alligator when flushed down the toilet. Conventional water treatments are able to break down a significant portion of the waste, though inevitably some would make it to the environment, continuing to potentially annihilate bacteria in rivers and lakes—both the harmful ones and the ones we like and need in our ecosystems. Even when it was broken down, triclosan yielded compounds just as nasty as the parent material. The two main degradation pathways were reaction with chloride ions to form 2,6-dichlorophenol (which lists instantaneous kidney failure on its dating profile and has a smaller LD50 than triclosan), and a UV-catalyzed degradation to yield a range of dioxins, a class of chemicals where an extra bridging oxygen is added between the rings, and which are highly toxic, wonderfully stable, and accumulate in the food chain.

But before you rush to the nearest sink and fearfully check the ingredient label of that bottle of liquid soap you've been using the last few weeks, terrified of seeing the words *triclosan* or *Irgasan*, there is a little good news. In 2016, almost 40 years after initial studies on a range of antibacterial additives, the FDA issued a list of 19 chemicals—including triclosan—that could no longer be incorporated into over-the-counter products intended for use with water. This went into effect in January 2017. Minnesota, where I was teaching at the time and still do, actually got ahead of this and passed a bill back in 2014 banning OTC triclosan products. Though that was something I should have celebrated, two days before the Minnesota announcement, I had just submitted a manuscript for publication that looked at a cool way of measuring triclosan in soaps based on a color-changing reaction. Thus, my timing sucked. But I guess not everyone gets to say their research was banned by the government, right?

And what about antibacterial soaps? Following the banning of triclosan, the replacement active ingredient in many of these products was benzalkonium chloride, which still sees widespread use at the time of writing in early 2023. While it's not anywhere near as nasty as triclosan, studies by the FDA exempted it from the 2017 ban due to insufficient data at the time. Something to think about the next time you wash your hands?

Since being a Doctor Who *nerd was not going to pay the bills, chemistry became the next best thing. Graeme grew up and did his first degrees in Scotland before coming over to the States for his PhD and never went back. Current life goal is making science accessible and relevant through teachings, labs, and community outreach.*

NERD NITE PITTSBURGH

LOST: Bladder Control. Reward for Safe Return.

by Maria Jantz

If there's one thing that would get most people pissed off, it's, well, not being able to piss. Unfortunately, in many neural disorders, including spinal cord injuries, multiple sclerosis, and Parkinson's disease, people lose the ability to control their bladders.

Scientists in pursuit of a particular form of trickle-down economics have figured out that we can use electricity to activate the neurons involved in bladder control when they're not working properly. This shocking discovery proved that stimulation could be applied in different locations and patterns to make the bladder either relax or contract.

So, urine trouble.

There are two kinds of problems that can occur in the bladder, although in a way both of them are number one. If you can't store urine (incontinence), you might find yourself taking drugs, and not the fun kind. If you can't pee on command, it is best treated by using catheters every time you need to pee. (Terrible as that sounds, it would be even worse if not for Benjamin Franklin. He invented the flexible catheter to help his brother with bladder problems—before that, people used straight metal tubing!)

If it sounds like these treatments would get on your nerves, you're right. That's why scientists have instead decided to get on your nerves with electricity to control the bladder. For a typical reflex to function, a nerve sends sensory information into the spinal cord, where it communicates with another nerve that activates a muscle. Fortunately, although a doctor might hit your knee to check some reflexes, they are unlikely to hit your junk with a hammer to test your bladder control.

This is the point at which you discover that I've been lying to you—in fact, no one has bladder control! You can't control your bladder directly; you can only control your urethra. Bladder reflexes start on the neurons coming from the urethra, which communicate with neurons in the spinal cord, which may communicate with neurons in the brain stem, and finally send a message to the bladder neurons. It's like a very high-risk game of telephone that may or may not end with you peeing your pants. Or to use medical terminology, voiding your pants—or certainly voiding the warranty.

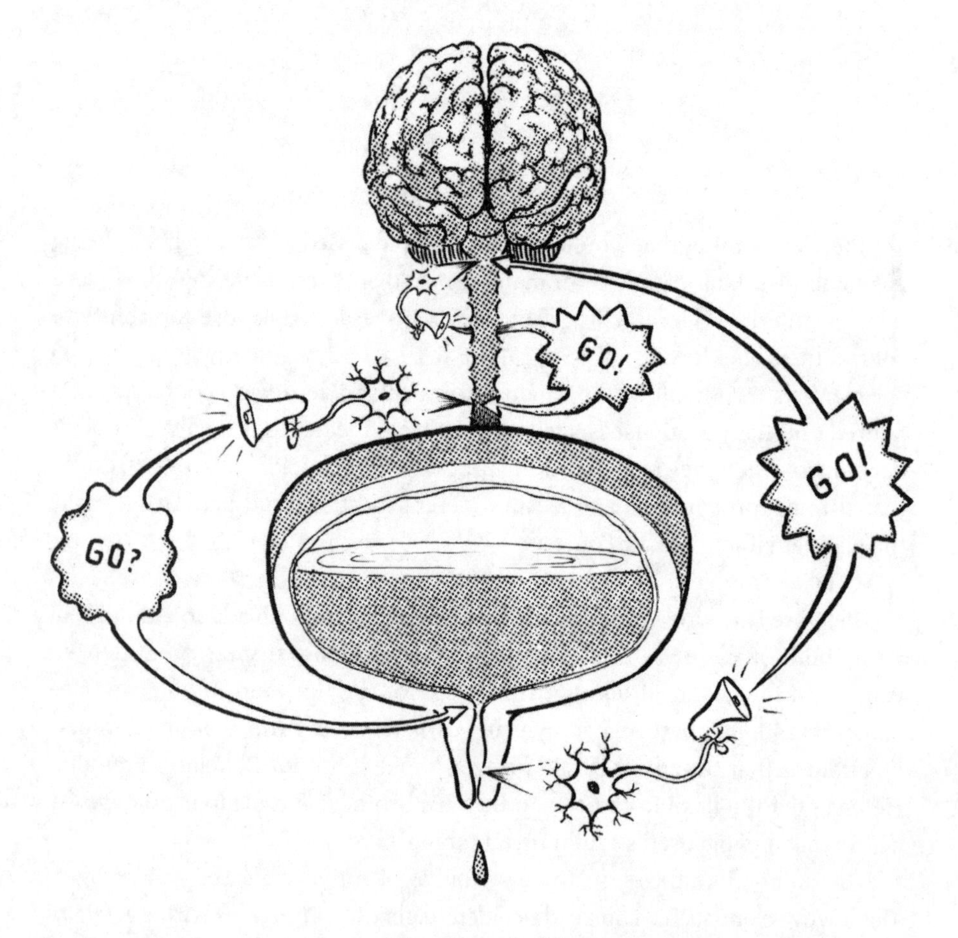

Surprisingly, one way to use electricity for bladder control involves applying electricity to your tibial nerve in your shin. This is clearly some distance away from the bladder, so what's happening? It turns out that some leg neurons interact with neurons in the spinal cord that project to the bladder in order to suppress bladder activity when you're walking around. If you've ever tried to walk and pee at the same time, you know it's a challenge (even more so than trying to walk and chew gum). Using electricity to increase activity on the tibial nerve can fool your system into reducing bladder activity.

Unfortunately, this only helps you if you pee too much, and not if you have trouble peeing. One simple way to induce voiding is by putting someone's hand in warm water at a sleepover. However, sleepovers are not very portable, so scientists have found other ways to make people pee, such as stimulating directly on the neurons from the urethra that initiate bladder reflexes. Interestingly, stimulation at this location is multipurpose, as it can either relax or contract the bladder. Very high-frequency stimulation pulses can cause voiding, while low-frequency stimulation on the same nerve can relax the bladder and restore continence. In this case, our ability to produce continence with stimulation relies on the reflex that makes you unlikely to pee during sex, which works because some of the nerves involved in bladder control overlap with those that convey genital sensation.

To pee or not to pee: That is the question the wisest sages, like myself, have pondered for hundreds of years. The answer, of course, is that one must do both. Life is but an endless cycle of filling your bladder, emptying it, refilling it, emptying it, and perhaps getting a nerve stimulator to continue this cycle. So get out there and embrace the anthem of *Frozen*: "Let it go, let it go, can't hold it back anymore."

Not content with her own urine, Maria Jantz has decided to embark upon the world's greatest pissing contest. Maria is a PhD candidate in bioengineering at the University of Pittsburgh. When she's not studying the bladder, she can be found making neuron mugs, doing handstands, or baking cookies.

NERD NITE GREENVILLE, NC

MICROBES CAN SAVE YOU, KILL YOU, OR JUST GIVE YOU THE POOPS

by **Ariane L. Peralta, PhD**

Microorganisms rule the world! They can be friends, foe, or acquaintance. Microbial friends that live in the gut can help digest foods we eat and transform and unlock nutrients that our bodies need. Microbial foes can hitchhike on some poop-contaminated foods and outcompete those microbial friends or acquaintances. And in some cases, poop-contaminated foods can transfer microbial foes into your gut, outcompete microbial friends, and give you the literal poops with a side of gut distress.

So how can we make sure to keep our microbial friends and acquaintances healthy while protecting ourselves from microbial foes? Wash your hands!

Even before the coronavirus hit the global scene in 2019, we were constantly reminded to wash our hands, but unless you were an employee serving food, this often seemed more like a suggestion. But it's always been a necessity. In fact, the US Centers for Disease Control and Prevention (CDC) even has a Handwashing in Community Settings section on its website that emphasizes how handwashing with soap *and* water saves lives. It says, "Keeping hands clean can prevent 1 in 3 diarrheal illnesses and 1 in 5 respiratory infections, such as the common cold or flu." The CDC and global partners even promote Global Handwashing Day on October 15 annually, and the World Health Organization (WHO) promoted World Hand Hygiene Day 2022 on May 5, 2022—its slogan was Unite for Safety: Clean Your Hands. Clearly, encouraging handwashing with products that clean hands is still a global challenge.

And if a website and a global day dedicated to handwashing wasn't enough, the WHO even has an entire campaign, WHO SAVE LIVES: Clean Your Hands, that highlights what you can do to nudge hand hygiene stats in the right direction. This is in addition to its public service announcements that implore people across the planet to "get involved in local hand hygiene campaigns and activities."

However, no matter how much the CDC or WHO implores us to wash our hands, a lot of our action or inaction, perhaps unsurprisingly, simply comes down to how much we think we're at risk of getting sick in a particular setting or situation. In research published in a 2020 edition of the *American Journal of Preventative Medicine*, Wändi Bruine de Bruin, PhD, and Daniel Bennett, PhD, found a direct relationship between initial COVID-19 risk perceptions and protective health behaviors, ultimately finding that our perception of risk is known to influence the likelihood of handwashing as well as other behaviors

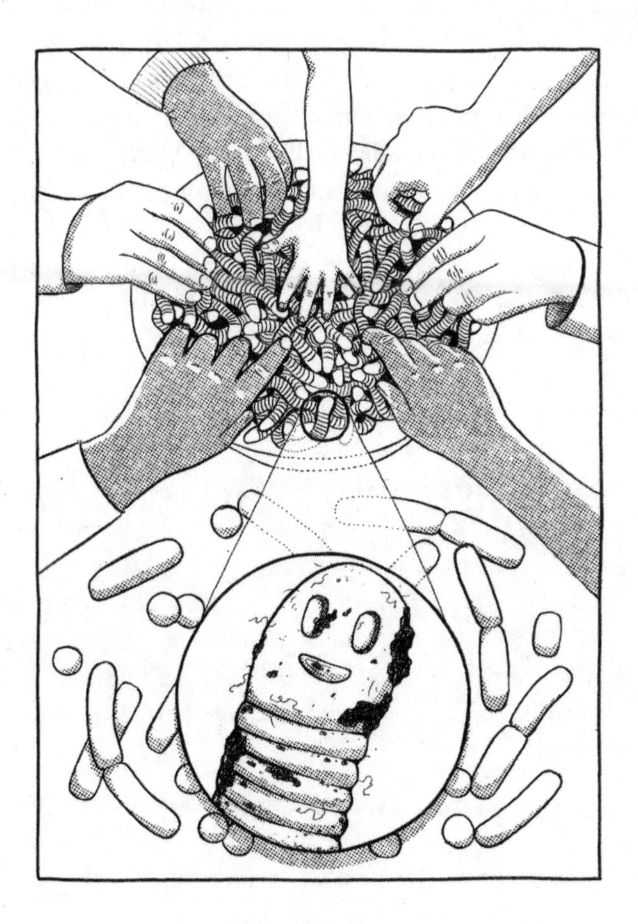

that protect ourselves, and one another, from spreading respiratory and diarrheal illnesses.

And while the public isn't always great about washing its hands, unfortunately, medical professionals aren't always on their p's and q's either. If you'd like to squirm a bit, read the charming 2022 study "Hand Hygiene Compliance in the Prevention of Hospital-Acquired Infections: A Systematic Review," which examined hand hygiene of health care workers and found that hand hygiene compliance varies widely and is challenging to monitor. Clearly there's room for improvement for our health care workers too.

So back to the rest of us. Even though literal signs everywhere remind folks to wash their hands—like in pretty much all public restrooms—our sanitation rating as a society isn't so great. I don't know about you, but I love a good after-dinner mint—except when the mints live in a bowl, are not covered, and are open to the elements. Even before the pandemic, the communal candy bowl was not a good idea. Oftentimes, these common-source candies are accompanied by a spoon that invites visitors to serve themselves. Thanks for the hollow gesture with that spoon. Let's be real. No one uses the spoon! People don't wash their hands after the bathroom and these same people take handfuls of candies from the bowl! But what about individually wrapped candies after dinner? This could be great for protecting public health, but more plastic is less good for the planet.

So where do we go from here? Wash your damn hands. And if you want to grab the mints or the candy-coated fennel seeds in that bowl, use the damn spoon and transfer those candies into your pre-washed hand. No more shitty mints!

Ariane Peralta is a professor who teaches biology and does research on microbes and learns for a living! AP wrangles environmental microbes to improve water quality, enhance wetland restoration, improve food production, and mitigate climate change. When not researching microbes, AP enjoys consuming microbial fermentation products, cooking and baking (microbes not always welcome), listening/watching/making music, and hanging with human and dog friends.

NERD NITE NYC

RUNOFF: What Is It and Why Even Notice It?

by **Evelyn M. Zornoza, RLA**

Roll up your sleeves, put on your work gloves, climb a ladder, and get up to your gutters. Oh, and please clean up after your dog does its business outside! Trust me, it's in your best interest.

Stormwater runoff is all the stuff that ends up in gutters—usually a combination of rain or melted snow or sleet, mixed with dirt, litter, and a wide range of pollutants. Runoff causes two major problems in urbanized environments. The first (and most obvious when it occurs) is *flooding*, which directly relates to the quantity of runoff and its timing. The second less visible—but still costly and dangerous—problem is *contamination* of our streams, rivers, lakes, and oceans.

Historically, engineers have come up with solutions for flood control in cities with varying levels of success. More recently, other professions (landscape architects, like me!) have contributed to the development of runoff solutions that address flooding but also tackle water quality *and* offer opportunities to create greener, more sustainable, resilient, and just plain better landscapes. (Am I biased? Maybe a little.)

And this is really why I'm running on about something as unappetizing as runoff.

Urban Stormwater Moves Fast and Gets Very Dirty (and Poopy)

To start, we need to look at what happens when it rains in a city—a raindrop's journey through New York, if you will. When rain falls on a city roof, a city sidewalk, a city parking lot, or a city street, the water moves downhill, across these smooth, often contaminated surfaces. Then it flows through a storm drain and into the city's sewer system, which delivers it to a wastewater treatment plant. Moving across slick, impervious surfaces like pavement and through smooth pipes, urban runoff is unencumbered by the porous soils, fibrous root systems, and soft surfaces that would otherwise absorb it and slow it down in natural, vegetated settings.

Besides moving too fast, the other problem with urban runoff is that it gets very dirty. In addition to picking up airborne contaminants as it condenses and forms in the air, runoff in cities travels over contaminated surfaces on its way to the storm drains. And in New York, as in all communities built before 1960, our sewer system is a *combined* stormwater and sanitary sewer system. This means that the runoff—even if it manages to make it across whatever impervious surface it travels without picking up any petroleum, heavy metals, salts, solvents, or other typical urban contaminants—never really stands a chance to remain clean, because once it enters the sewer system, it is destined to mix with—yes, you guessed it: poop. Because, well, poop is pretty much everywhere outside.

So at this point in its journey, our original raindrop is incorporated into a foul, polluted slurry, speeding through underground pipes toward a wastewater treatment facility.

Luckily, on most dry, sunny days and even days with small rain showers spread across a long time period, our wastewater treatment facilities work pretty well and can process the stormwater and our sanitary waste before they release it, treated and cleaned, into an adjacent river, lake, or stream. The key runoff problems (flooding and contamination) arise when the storm is big and the rain falls suddenly. When large quantities of stormwater runoff are generated by big storms quickly, storm sewers can get backed up, which is when streets and basements flood. In addition, large storm events also overwhelm the treatment facilities, which will then release untreated stormwater into the surrounding water bodies. These untreated releases include the runoff that likely picked up debris, motor oil, salt, and detergents before it got to the sewer—plus the poop, cooking grease, and everything else New Yorkers put down their drains. This is what is called a combined sewer overflow (CSO) and is the reason swimmers, surfers, and beachgoers are often greeted by signs warning them to stay out of the water on the days following large rainstorms.

If we want to reduce the impact of urban runoff, particularly CSOs, the key is reducing the peak of the hydrograph, a measurement of the amount of runoff, especially when it exceeds the capacity of the stormwater system. To better understand this, I'm excited to introduce you to . . . the rational formula! Engineers and hydrologists have used this formula for a long time to predict peak runoff (the top of the hydrograph curve) for a given storm event. The formula is Q = CiA, where *C* is a runoff coefficient, *i* is the rainfall intensity, and *A* is the area of the watershed. In most cases, designers cannot control i or A. We cannot control how hard it rains and we are, for the most part, stuck with our watershed boundaries, which have been determined by either geography or previous generations of engineers. But in the past few decades, landscape architects and environmental engineers have developed alternative landscape solutions that reduce the urban hydrograph's peak by changing C, the runoff coefficient.

Reducing C and Beautifying Your City: Keeping the Feces Where It Was Intended to Be

What is a runoff coefficient? It's just a number that represents how smooth and nonabsorbent or how rough and spongy a material is. (Sometimes it's just called a roughness coefficient.) Landscape solutions to runoff and stormwater control attempt to reduce C in the rational formula (and ultimately, Q) by making the surfaces associated with our stormwater collection and conveyance systems rougher and more absorbent. Designers have created surfaces that mimic landscapes with runoff coefficients somewhere between Forest (~0.2) and Turf (~0.4). Here are some examples that your city should try:

Swales: Swales are linear, vegetated channels whose main purpose is to convey runoff (like a pipe) while also absorbing and filtering out some of the contaminants. Swales in urban areas are usually designed to accommodate relatively small runoff volumes (not a hundred-year storm). This is because the runoff from the first ten minutes of a storm tends to have the most concentration of pollutants. This means that the runoff reaching treatment facilities during a typical storm will have fewer contaminants.

Bioretention basins: Conventional bioretention basins often require more space than we have available in most New York City neighborhoods, but designers have adapted some smaller-scale rain garden and tree-pit designs to incorporate bioretention functions. Bioretention basins can hold larger quantities of runoff than a swale, allowing some of the water to be absorbed and releasing the rest into a sewer system after a storm's hydrograph peak has passed.

Green roofs: Hydrologically, green roofs function similarly to bioretention basins. They are just much flatter and aren't installed on the ground. They provide a reservoir layer and vegetation, both of which retain and filter the water. Like retention basins, they reduce C and delay the release of the water into the conventional sewer system.

Porous paving materials: Sometimes paving is necessary. Unreinforced, vegetated surfaces cannot support vehicles or even provide universal access for people with mobility limitations. But not every paved surface has to be solid asphalt or concrete. Unit pavers, designed with water storage reservoirs as part of their setting beds, can replace traditional impervious paving in some applications.

While these types of stormwater infrastructure reduce the pollutants entering our waterways, they have an added benefit by inserting planting—ideally, native vegetation—into what would otherwise be inhospitable swaths of paving or roofing, or miles of pipes. This is a key point, because the biggest obstacle to implementing some of these components is that they cost money (both installation and ongoing maintenance) and take up space. The fact that they can also be both beautiful and restorative beyond the purview of runoff control should be included in their cost/benefit analysis.

Evelyn Zornoza is a licensed landscape architect with twenty years of experience on both residential and public projects. She has Master of City Planning and Master of Landscape Architecture degrees from UC Berkeley.

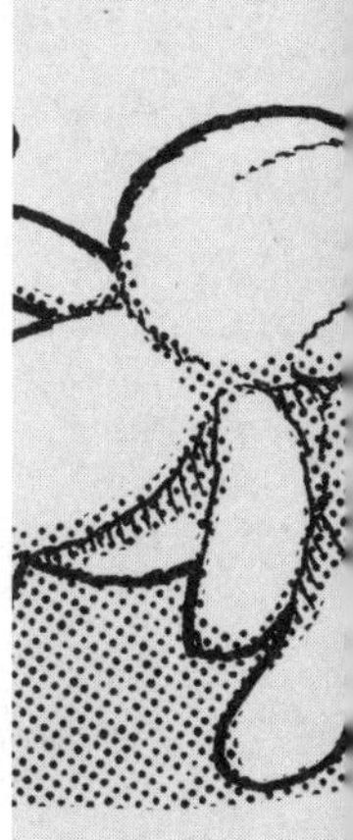

Doing It

Birds do it. Bees do it. Even educated fleas do it. You know what else does *it* (and in this case, unlike Cole Porter in 1928, we aren't talking about falling in love)?: roadrunners, coyotes, sheepdogs, voice-over artists, illustrators, wascally wabbits, the Fudds, starfish and paramecia (by themselves), adult entertainment performers (amateurs and professionals), parents, grandparents, siblings (ideally not with each other), cousins (second or third ideally?), cows (except steer), horses (except geldings), cats (except for most indoor cats—my cat, Lu, will die a virgin), partridges in pear trees, geese, swans even when a-swimming, frogs, and pigs. Also, mammals in savannas (some of which do it sneakily), ducks (yes, we all now know that male ducks have corkscrew penises), the young siblings in *Flowers in the Attic* (ewww!), the hawks that frequently roost on a lamppost in my neighborhood in Southern California and the rooster that lives up the hill from me, and seemingly every character on every HBO show, do it. And probably most of you.

As George Michael once crooned, "Sex is natural, sex is good." At its most fundamental, sex propagates species. It moves the world forward. It leads to evolution. It leads to, well, basically everything. And it can vary wildly among species and even within a single species. So let's get into *it*. Warning: This chapter may make a Puritan blush.

—Matt

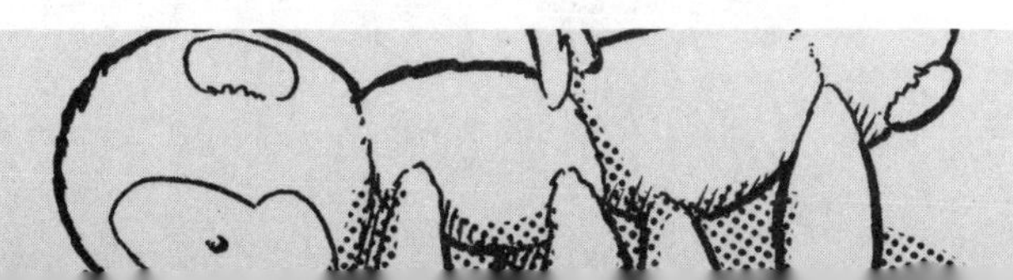

4K

NERD NITE NYC

HOT OR NOT? HOW TO BE A PERFECT 10; or, How to Manipulate Perceptions of Physical Attractiveness

by **Lillian Park, PhD**

Who hasn't thought about what it would be like to be hotter? While you might be happy with your current level of physical attractiveness, you might not mind if one day you woke up and bam!, you're a perfect 10. Beyond plastic surgery and a ton of makeup, what could you do to change how attractive people find you? Here are five quick tips on how to manipulate how people perceive your physical attractiveness.

Wear Red

The color you wear could have a significant impact on how hot you appear. Researchers have found that wearing the color red increases how physically attractive people find you. It doesn't even need to be an entire article of clothing. A red necktie or red lipstick will do the trick, too!

Have Good-Looking Friends

This may seem counterintuitive, but it turns out that good-looking friends help us by elevating perceptions of our own physical attractiveness. Yes, being surrounded by good-looking friends makes you appear better-looking than if you were alone or surrounded by plain-looking friends. Researchers at Michigan State University found that having attractive friends on social media led people to thinking the profile owner was more attractive. So let your gorgeous friends do the heavy lifting. Surround yourself with them and voilà, your own attractiveness will be elevated.

Let People Approach You

Is it better to be the hunter or the hunted? While it's tempting to think that being the hunter gives you the upper hand, when it comes to hotness, it turns out you want to be the hunted. Researchers from Northwestern University conducted a study on romantic attraction and selectivity at a speed-dating event. Half the time, the men moved from table to table, while women sat at the same table. The other half, the women moved from table to table, while the men sat at the same table. Surprisingly what mattered in determining attraction was who was doing the moving! The people who moved experienced greater attraction and chemistry toward their speed-dating partner. Next time you see a hottie, make them come to you!

Wait Until Closing Time

It's true, people do become more attractive at closing time. You might be thinking, beer goggles, right? Well, that's not quite the whole story. Researchers in Australia conducted a study in a pub in Sydney in which bar patrons were asked to rate other patrons at the beginning of the evening, the middle of the evening, and closing. They also breathed into a Breathalyzer to measure blood alcohol concentration. While it is true that people got drunker as the night progressed, that alone couldn't account for changes in perceptions of attractiveness. Researchers concluded that repeated exposure and the diminishing pool of potential romantic partners throughout the night led bar patrons to revise their perceptions of the attractiveness of the remaining people at the bar. Thus, go to a bar, let yourself be seen, and wait until closing time when there's not as much competition. Easy!

Personality *Does* Matter

Human beings aren't as shallow as we think. In fact, your physical attractiveness can be manipulated by your personality. There was a study where people rated physical attractiveness of people in photos. Half the photos were accompanied by desirable personality traits, such as honesty, intelligence, humor, and kindness, whereas the other photos had negative personality traits, such as offensiveness, cruelty, rudeness, and abusiveness. Thankfully, people found the photos with the desirable personality traits more attractive. Remember the previous study where being surrounded by good-looking people increased your attractiveness? In that same study, researchers found that the comments and behaviors from the good-looking friends mattered. Positive comments and kind behaviors from the good-looking friends increased attractiveness in social media profile owners. We want to be with people who are kind, generous, dependable, and supportive. There is nothing sexier than someone who makes us feel good. Being the kind of person that other people want to be around because of your sterling personality immediately boosts your hotness.

Remember, the next time you go out, follow these five quick tips and be the charming stranger in red surrounded by attractive friends that people approach at closing time. So hot!

References

Elliot, A. J., and D. Niesta. "Romantic Red: Red Enhances Men's Attraction to Women." *Journal of Personality and Social Psychology* 95, no. 5 (2008): 1150–64. https://doi.org/10.1037/0022-3514.95.5.1150.

Elliot, A. J., et al. "Red, Rank, and Romance in Women Viewing Men." *Journal of Experimental Psychology: General* 139, no. 3 (2010): 399–417. https://doi.org/10.1037/a0019689.

Finkel, E. J., and P. W. Eastwick. "Arbitrary Social Norms Influence Sex Differences in Romantic Selectivity." *Psychological Science* 20, no. 10 (2009): 1290–95. https://doi.org/10.1111/j.1467-9280.2009.02439.x.

Johnco, C., L. Wheeler, and A. Taylor. "They Do Get Prettier at Closing Time: A Repeated Measures Study of the Closing-Time Effect and Alcohol." *Social Influence* 5, no. 4 (2010): 261–71. https://doi.org/10.1080/15534510.2010.487650.

Lewandowski, G. W., A. Aron, and J. Gee. "Personality Goes a Long Way: The Malleability of Opposite-Sex Physical Attractiveness." *Personal Relationships* 14, no. 4 (2007): 251–585. https://doi.org/10.1111/j.1475-6811.2007.00172.x.

Walther, J. B., et al. "The Role of Friends' Appearance and Behavior on Evaluations of Individuals on Facebook: Are We Known by the Company We Keep?" *Human Communication Research* 34, no. 1 (2008): 28–49. https://doi.org/10.1111/j.1468-2958.2007.00312.x.

Lillian Park, PhD, received her doctorate in psychology from the University of California–Berkeley. She is currently an associate professor and the chair of the Department of Psychology at the State University of New York–Old Westbury.

NERD NITE PHOENIX

DATING TIPS FROM THE ANIMAL KINGDOM: What to Wear and How to Flaunt It

by Kaci Fankhauser

Any freshman marketing major can tell you: Sex sells. Whether it's a magazine cover in a grocery store or sponsored content on your news feed, marketers rely on visual media to grab our attention. You might see some oiled-up abs paired with a canned refreshment (talk about a thirst trap) or a tanned bikini babe seductively biting into a trendy new burger—and whether we want to be them or get with them, a hottie in a product advertisement is pretty good at convincing us that we need what they've got. We humans also use visual cues to advertise ourselves—I might wear a fitted blazer to convey confidence at a business meeting, then change into something a little more salacious to post a selfie I'm hoping my crush will see on Instagram. But in the absence of gym selfies encouraging a potential mate to swipe right—how do members of the animal kingdom sell sex?

Turns out, non-human pickup artists have quite a few tricks up their (metaphorical) sleeves.

Some species have developed seductive choreography to woo a mate, while others express their readiness for copulation in song. Much like in the human world of marketing, mate selection in the animal kingdom often relies on visual cues. If you've ever spent hours before a first date doing your hair and picking out an outfit, you know that some sleek plumage or an impeccable pop of color can really elevate a first impression.

Let's talk about that pop of color for a minute. On first pass, the concept of a color seems simple. ROY G BIV has us covered. But, as with all life, it's a bit more complicated than that. Let's revisit some high school science, shall we?

What we see as color consists of waves of light, and we perceive different-sized waves as different colors. So light with a wavelength (the distance between two peaks of the wave) of 700 nanometers looks red, while light with a wavelength of 400 nm looks violet. An object looks to be a certain color based on the wavelengths of light that bounce off that object—so a red ball reflects light in the seven-hundred-nanometer range, which our squishy attics (brains) process as the color red.

For scientists who study animal coloration, it can be a challenge to define the colors themselves. What one scientist calls "aqua" another might call "teal" and yet another, like one researcher with color-blindness I know, might call "gray." To get around trying to decide if something is more of a fuchsia or a magenta, many scientists use a tool to measure color called a spectrophotometer. It works by shining a bright white light containing equal portions of all visible light at an object (think of the white light from Pink Floyd's *Dark Side of the Moon* album art) and then measures the amount of each wavelength of light that is reflected off the object. So if you were to measure a red ball with a spectrophotometer, the resulting graph would show you a higher reflection of wavelengths around seven hundred nm and little or no reflection of light in other wavelengths.

And once we have that graph—called a reflectance spectra—we can do some really neat categorization of colors. First we can define what the color is by wavelength—that's the hue—and then we can measure brightness, which is the height of the reflection peak, and chroma, which is essentially how "pure" the color is. If a reflectance peak at 700 nm is tall and skinny, we will see the color as bright and red; but if that peak is short and wide, showing more reflection in the 600 range, we would see more of a dull red-orange.

Okay, back to the animals with their fancy coloration. The donning of bright colors in the animal kingdom is key for first impressions—especially for mate choice. Not all bright colors are used to win a mate; sometimes bright colors on a critter signal danger, like the neon hues of a poison dart frog. But what sets the coloration used in mate choice—called sexual signals—apart is that often it is only present or significantly exaggerated in individuals of one sex. This is called sexual dimorphism—*di* meaning "two" and *morph* meaning "type." You probably can think of several animals that display sexual dimorphism—male peacocks, for example, have big, colorful tail-fan displays, while female peahens are mostly shades of brown. Additionally, female black widows are large and black with the notable red hourglass, while males are small and brown. Male lions have larger manes than female lions, and male deer have larger antlers.

Why do we see these differences in physical traits between sexes of a species? These differences—the sexual signals—influence mate choice via sexual selection. Colors used in sexual signals aren't just about aesthetics. Often an animal's coloration can tell a potential partner about an individual's quality as a mate.

Take the pipevine swallowtail, *Battus philenor*, for example. This butterfly, native to much of the United States and Mexico, is black with iridescent blue hindwings. And while both males and females have this coloration, the blue on males' wings is significantly brighter and, well, bluer. But scientists have discovered that the wings among males aren't uniformly bright and colorful. Male pipevine swallowtails with wings that are a brighter blue bestow upon their mates larger spermatophores—"nuptial gifts"—of sperm and nutrients transferred to the female during sex. So a female that "don't want no scrub" might reject the advances of a duller-looking male.

When it comes to sexual signals, we can also take a lesson from the animal kingdom on what to wear *and* how to flaunt it. On a warm summer morning in the Arizona desert, you might be lucky enough to come across male pipevine swallowtails perched on a tree with their backs to the sun, angling their wings *just so* in order to make their iridescent blue shine as brightly as possible. It's like when you're trying to take a cute selfie—lighting is key. This isn't so different from what we humans do to show off our favorite features—many of us have a go-to photo pose to highlight our best angles. And on the other side of the globe in an Australian rain forest, a bird of paradise with his brightly colored plumage will spread out his feathers in front of a female and shake what his mama gave him—a cloacal choreography. It's basically bird twerking.

So next time you're getting ready for that first date, take a page from the animal kingdom look book. Whether you go for a flashy shirt or a voluptuous updo, just channel the confidence of a male red-capped manakin doing his best impression of a moonwalk to impress the ladies.

Although this essay describes simplified, heteronormative examples of animal courtship, the author would like to emphasize that these examples do not fully encompass the diversity of sexual expression in the animal world. For example, take a gander at the chapters "Finding Nemo('s Sex): Sex Change and Gender Roles in Anemonefishes," if you inexplicably skipped that one, and "Going Ape for Pansexual Primates."

Kaci was born and raised in Phoenix, Arizona, and fell into doing research accidentally for the chance to just hang out with some butterflies. She attended Arizona State University and Northern Arizona University and earned degrees in biology (BS), anthropology (minor), and environmental science and policy (MS), where she especially loved learning about the intersection of evolution and ecology. For the past six years, she has worked for SciTech Institute, a STEM education nonprofit, where she strives to make science a field that is accessible to and makes the world a better place for all *people.*

NERD NITE DC

DATING AS A DATA NERD

by Tristan Attwood

Part One—I Should Start Dating

So this whole thing started in mid-2012. One afternoon I was hanging out, taking photos of my cat (as you do), and when I went to send one particularly good one to a friend, I realized that almost all of my messages to them were pictures of my cat. I then looked at my messages to a different friend and saw the same thing. And then a third friend. My whole camera roll was cat pictures. Now, I'm an analyst by trade, but I didn't need a sophisticated algorithm to realize that maybe I needed to have more going on in my life. With this realization under my belt, I decided that the logical thing to do was to try dating.

A brief aside here. The online dating landscape has changed dramatically since 2012, but at the time the biggest name in dating was OkCupid. Tinder wasn't a thing yet, and there were a few other options, but OkCupid had an amazing blog on the data science it used to match people—a concept that deeply appealed to me and my background. So OkCupid it was.

The way it worked back then was that you made a profile, uploaded a picture, entered some basic information about yourself, and started answering questions. The site then used your responses to calculate a score ranging from 0 to 100 percent, letting you know how compatible the algorithm thought you were with any specific person.

I started matching with people and going on dates. At first, it was a lot of fun.

Nothing serious came of them, but they gave me confidence and everything seemed to be going well. Then I matched with her.

A few physical details are important here. I'm six feet tall and fairly broad. She's five feet and very slender. I only mention this because that disparity is important context for what you're about to read next.

We matched, chatted a bit, and then decided to grab drinks at a local happy hour. Most of our conversation was pleasant but unremarkable first-date material. Then out of nowhere, she suddenly announced that she "doesn't believe in rape." I was caught wildly off guard and said only, "Oh?" She responded by saying that, in her mind, a woman could always fight someone off. Against my better judgment, I argued with her about it, bringing up the many reasons this is simply not true. Her response? That she had a combat knife in her purse and knew how to use it. At that point, I changed the subject and made some bland chitchat until I could leave. I gave her a fake phone number, went home, and fell all the way into my own head about things.

The algorithm had said we were 99 percent compatible. What did that say about me? Do I also hold these sorts of survival-of-the-fittest, stab-or-be-stabbed beliefs and not know it? Algorithms had a lot of sway over people in 2012.

After thinking about that for a little too long, I decided that I needed to know, and I knew how to find out. I needed to start building my own algorithm and find out for myself.

Part Two—Data Will Solve This

Now that I had decided to data my way out of this, I actually needed to collect data. So I did what every good analyst does when they have a project to do and don't want to work too hard: I opened Excel. I eventually settled on collecting the following fields as a starting place:

- Name
- Age
- Occupation
- Siblings
- Degrees
- Background
- My impressions/notes

After collecting this data from a couple dozen dates, I started looking at it and just trying to see what popped out as an interesting pattern. And luckily, I found one!

OkCupid did not show you any real first names back in 2012, just screen names. I had no idea what anyone's real name was until we met. Now, according to the US Social Security Administration, the most common name for women in the age range I was dating is Jessica, with a little over 20 percent of women having it as their first name. Conversely, Katherine is much farther down the list, with approximately 5 percent of women having it as a name. But when I looked at my data, a full 40 percent of people were named Katherine or some variant of it (Kate, C/Kathy, C/Kat, et cetera). I didn't know how knowing that my next date would be named Katherine helped me, but it really cheered me up and let me know that I was on the right track and was able to start pulling interesting insights from the data. The issue, I decided, was that I wasn't collecting *enough* data.

Part Three—*More* Data Will Solve This

At this point I took an inventory of what I was doing and realized I needed to put in the effort and build a proper database to house this. Excel had done great for its part, but it was time to get serious. So naturally, I built a proper relational database and decided to add the following fields to the ones I already had:

- Religion
- Importance of religion to them
- Ethnicity
- Nationality
- Dietary preferences
- Hair color
- Body type
- Height
- Sense of humor
- What are they proud of
- Who initiated contact
- . . . a few others

This effort really paid off—but not at the beginning!

The problem I quickly encountered was that I entered into most dates having data collection at the top of my mind, which basically means that I was a *terrible* date, as it's unsurprisingly hard to collect 20-plus individual bits of information about someone while seeming natural and engaging in normal, fun, charismatic first-date conversation. Luckily, I realized this (fairly) early and was able to course-correct before I got too creepy.

In looking at this data, a clear pattern started to emerge. From my data, I learned that they were likely to connect with me if they were, among other factors, ethnically Jewish, did not consider religion a big part of their life, were within a year of my age, were an only child, had a dark sense of humor, and attended a small liberal arts college. And this was fantastic for a number of reasons. First, people I really connected with were not, contrary to some other algorithms, all knife-wielding social Darwinists. Second, it was a great personal validation where I had set out to use data to answer this question and had successfully done it. And third, over the course of this I ended up going on over 250 first dates. Now, the vast majority of these weren't interesting or fun or sexy, they were just 20 minutes of no-chemistry conversation at a bar or coffee shop before we both said goodbye and never saw each other again ever. And when you go on that many failed dates, it really starts wearing on you emotionally. But with this profile I had constructed, I was able to be much more selective about both dates and my emotional energy. This selectivity was key, because it let me have the emotional energy to realize how great the woman who was about to message me was.

Part Four—Suddenly Jessica

One day out of the blue, a woman named Jessica messaged me. After chatting for a bit, two things became crystal clear:

> She's awesome.
>
> She matched 13 of the 15 factors that I had been able to identify as being predictors of things going well. The only two missing were that she had two siblings and was not ethnically Jewish.

And because I had that 13-out-of-15 figure, I was able to put a *lot* of my emotional energy into her and the relationship. This absolutely paid off a few years later when we got married, and a few years after that when our son, Roland, was born. But here is the punch line to this whole thing. I had gone through all these dates, built databases, and run countless analyses to get to this person. Jessica had signed up for OkCupid, looked around, and *messaged me*. And . . .

I was her first-ever online date.

> *Tristan Attwood is an analyst working away in the data mines for Big Airplane. Happily married to his lovely wife, Jessica, he spends most of his time attempting to keep up with his young son.*

NERD NITE ORLANDO

10 THINGS YOU DIDN'T KNOW ABOUT SEX . . . EDUCATION

by **Anna V. Eskamani**

You're an adult, so you totally know all there is to know about sexual health and prevention, right?

Wrong! There is so much information out there, and some things we might take for granted, too. Here are some of the biggest sex-ed rumors; learn the facts and get ready for some fun surprises. Isn't science cool?

Okay, sex-ed rumor number 1: There is no such thing as safe sex.

Reality: This is a super common myth, especially in areas where abstinence-only sex education is prevalent. While, yes, the only way to 100 percent guarantee that you will not become pregnant or contract an STI is to remain abstinent, there are plenty of ways to be safe when it comes to sexual activity.

Sex-ed rumor number 2: You can't get pregnant the first time.

Reality: Anytime you engage in sexual activity—even when you are using protection—there is a chance of pregnancy.

Sex-ed rumor number 3: Jumping jacks after sexual activity prevents pregnancy.

Reality: Yay for you! You're very fit and active but *no*—this will not prevent pregnancy. The only thing you'll do now is increase your heart rate. Of course there are effective ways to prevent pregnancy, but this isn't one of them.

Sex-ed rumor number 4: Coke is a great spermicide.

Reality: Coke has many unconventional uses. You can soak pennies in it to make them shiny. You can use it to tenderize a pot roast. You can even use it to help get sticky gum out of your hair. Coke does not, however, work as birth control.

Sex-ed rumor number 5: Pulling out is an effective form of contraception.

Reality: It is not, and pregnancy can still take place and STIs can be contracted, too. Best bet is to use a barrier method of contraception, like a condom.

Sex-ed rumor number 6: Masturbation is bad for you.

Reality: Masturbation is very healthy and safe. Why is the rumor that it's bad for you still floating around?

Sex-ed rumor number 7: Having protection available means your partner will want sex.

Reality: This is not true—you must always look for consent, and despite past teachings of "no means no," what consent actually means is looking for an enthusiastic yes. A person can also take away consent at any time.

Sex-ed rumor number 8: Using two condoms is better than one.

Reality: Condoms are incredibly effective in preventing pregnancy and the transmission of STIs. However, using two condoms at once can actually lead to breakage and could damage the protection you need.

Sex-ed rumor number 9: Abstinence-only education is effective in reducing unintended pregnancy and STI rates.

Reality: This is a big myth—there is no evidence to back any claim that abstinence-only education works.

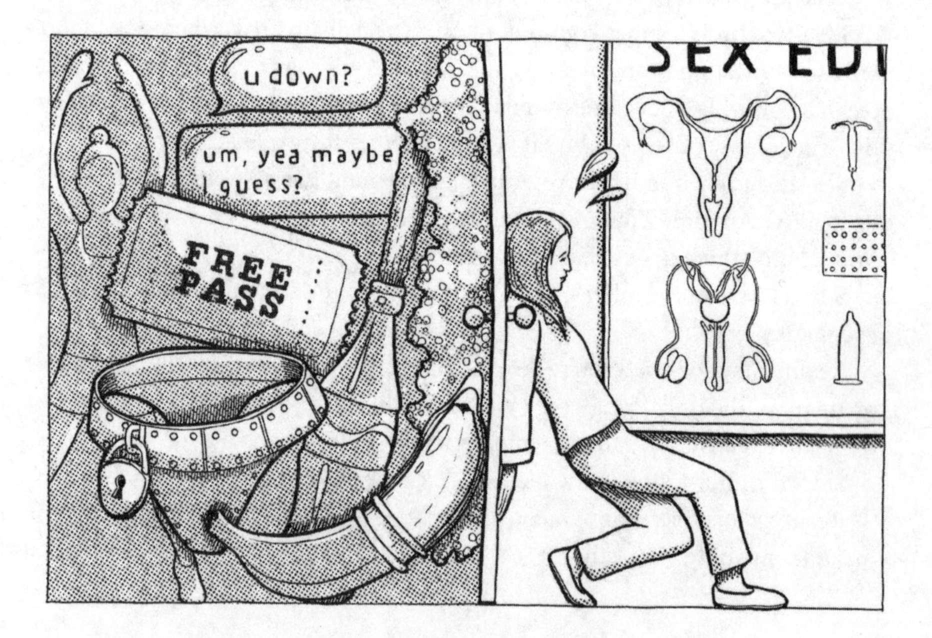

Sex-ed rumor number 10: Comprehensive sex ed encourages youth to have sex.

Reality: Another big myth! In fact, research tells us that when individuals know the risks of sexual activity and the means to stay safe, they are more likely to abstain and act responsibly if they do engage. Knowledge is power!

Anna V. Eskamani is an Orlando local, lifelong advocate for reproductive rights, PhD student at the University of Central Florida and the state house representative for Florida's 42nd District. When she's not working, she's watching Star Wars.

GOING APE FOR PANSEXUAL PRIMATES

by Natalia Reagan

Happy Pride to everyone who's celebrating! However, it's not just humans that are partying in the streets—there are over 1,500 other species celebrating in the jungle, seas, tundra, and deserts. Yep, that's how many species scientists have documented that engage in homosexual behavior.

And to be clear, just because we witness an animal engaging in homosexual behavior doesn't mean researchers immediately label the animal as gay. It is rare to document strictly homosexual behavior in the course of, say, one penguin's life. Who knows what or who Steve is doing when we're not watching? (Read: We haven't found any self-identified Kinsey Scale 6s . . . yet!)

Also, in the past many scientists have tried to explain away the gay in many species saying that the homosexual behavior is less about preference or pleasure and more about achieving dominance.

Right. Okay. Suuuuuuuure.

This is also why it is important to have a wide variety of scientists from different backgrounds with a wide range of perspectives in all scientific fields.

But I digress.

Let's talk about pansexual primates. What do we mean by pansexual? "Pansexual is sexual, romantic, or emotional attraction toward people regardless of their sex or gender identity." Now, because we can infer emotional attraction in primates, we will be basing our criteria on homosexual behavior.

What do we mean by homosexual behavior? Sexual activity between members of the same sex, not always anal or vaginal penetration. Mounting, crural sex (a fancy way to say "hand stuff"), genital stimulation, mutual masturbation, oral sex, and genital presentation are all included in the umbrella term *homosexual behavior*.

Now that we have our criteria, let's meet the top 3 queerest primates!

3. Macaque Attack

Coming in at number 3 are macaques, specifically Japanese and stump-tailed macaques. When people ask me if there could be a planet of the apes, I like to remind them, yes, we're already living in it since humans themselves are great apes. *But* if we should die out, who would take over? Well, if it isn't the cephalopods (see the chapter: Cephalopods: The Impossibly Awesome Invertebrates), it'll definitely be MACAQUE ATTACK! That's because they—like humans—are able to survive in nearly any environment, including deserts, urban cityscapes, and snowy mountaintops. Macaques are very plastic and flexible in their behavior—especially sexual behavior.

There are more than 20 species of macaques, and they're all over the sexual spectrum. For example, female Japanese macaques will form consortships with other females for extensive periods of time, and scientists have documented some females that prefer the company of other ladies. And some Japanese macaque males aren't so picky, either. In fact, they're known for their intense interspecies mingling, specifically mounting sika deer.

Stump-tailed macaques males will engage in all sorts of homosexual behavior, including mounting and orally stimulating one another. Some of their positions would make the authors of the *Kama Sutra* blush. One such position is essentially a monkey 69 with a two-fer-one fellatio special.

2. Gay-ish Gorillas

Coming in at numero dos are gay-ish gorillas. Gorillas have been mischaracterized and misunderstood for centuries. They're more gentle gay giants than terrifying tyrants. Sure males can be competitive when it comes to the ladies, but sometimes gorilla males like to forgo the gals and get frisky with each other.

In 2018, two three-year-old western lowland gorillas named Aybo and Thabo gave Rotterdam Zoo visitors quite a show with their agile amorous pursuits. These juveniles were seen in several seductive positions—including a rarely spotted missionary moment (missionary is most often associated with unadventurous humans)—and at another point, one male bent the other one over and gave the most Yes-I've-been-caught-but-I-still-don't-give-a-Rotterdam look to the camera I've ever seen.

But these were captive gorillas—what about gorillas in the mist? Well, in one study at Volcanoes National Park in Rwanda, University of Western Australia researchers witnessed 18 out of the 22 mountain gorilla females they were studying in *ahem* compromising positions. This included genital closeness, genital rubbing, and specific vocalizations—aka gorilla dirty talk. These observations directly challenge outdated assertions that homosexual behavior in primates was just about dominance. These gorilla gal pals were simply enjoying themselves. A lot.

Scientists also notice that on a few occasions this "girlrilla-on-girlrilla" action happened after they had witnessed sexual behavior between a male and female in the group. Researchers surmise this might have had a pornographic effect and that, once these gorilla gals got riled up, they opted to use each other as a sexual outlet.

1. Bone-obos

Coming in *hot* at number one are bonobos (*Pan paniscus* or Pan promiscuous). Bonobos and chimpanzees are our closest genetic relatives, sharing nearly 99 percent of their DNA with humans. However, they handle conflict *very* differently. Chimps often get aggressive, while bonobos have a more diplomatic way to defuse tension. How?

Boning, boffing, schtupping, doing squat thrusts in the ol' cucumber patch.

Yes, bonobos engage in sexual activity to defuse tension. So instead of a strongly worded email or a fistfight, these great apes simply work out their issues on the forest floor . . . or tree . . . or river. Eh, you get the idea.

And it's not just heterosexual play—*anything* goes with bonobos. And it has been found that females that engage in more frequent acts with members of the same sex are higher on the proverbial ladder of bonobo babes. Genito-genital rubbing is often seen between bonobo females—think primate scissoring. They also love to cuddle and kiss. With tongue.

Male bonobos are not exempt from this lustful sex fest. They will engage in penis fencing, which is the actual scientific term for the more colloquial *sword fights* (you're welcome). And when males are reunited after some time, they often engage in scrotum rubbing, where one male will rub his family jewels on the buttocks of another.

Research has shown that nearly 75 percent of all sex between bonobos is nonreproductive. Essentially all bonobos are what we would consider bi- or pansexual.

Now, before you start the "Bonobo Conflict Resolution" campaign at your place of employment, you should know something else about bonobos. They are female-dominant. Yes, bonobos live in matriarchal societies, and the ladies run the show. So if you want to be more like bonobos, perhaps you should forget her emails, quit with the mansplaining, and start voting more women into office.

Born This Gay

There are over 1,500 species that engage in regular, natural homosexual behavior, and there is still one species known for propagating homophobia—*Homo sapiens*. Get it together, humans!

Macaques, gorillas, and bonobos are merely three examples of pansexual primates—there are many, many more! If you'd like to know more about primate sexuality, I suggest digging up a copy of *Primate Sexuality* by Alan F. Dixson (yes, that's his real name). The illustrations alone are worth it. And I implore you to go down a gay animal rabbit hole (including the eastern cottontail rabbit) and discover that sexuality—like most things in life—is absolutely fluid, flexible, and fabulous!

Natalia Reagan is good at two things: (1) making science funny and (2) chasing monkeys. She is an anthropologist, primatologist, comedian, host, producer, podcaster, professor, writer, and monkey-chasing weirdo. Her training was in a four-field approach to anthropology, and she often examines topics through a bio-cultural lens. Natalia was a comedy writer and correspondent on Neil deGrasse Tyson's StarTalk, *a regular host of the* StarTalk All-Stars *podcast, an animal expert on Nat Geo Wild's* Everything You Didn't Know About Animals, *a science expert on History's* UnXplained, *a skeptic on Travel Channel's* Paranormal Caught on Camera, *and the co-host of Spike TV's* 10 Million Dollar Bigfoot Bounty. *For her master's fieldwork, she conducted a survey of the Azuero spider monkey in rural Panama. Her passion includes combining science and comedy to spread science literacy while inducing spit takes.*

NERD NITE MADISON

SMELLS AND THE MICROBIOME: Are Microbes Controlling Your Sex Life?

by Dr. Jenny Bratburd

Despite what you may have heard, humans are not bad at distinguishing smells. Like other mammals, we've got pretty decent olfactory systems with somewhere around 400 receptors to detect smells. But differentiating those smells requires training.

While we often get some kind of formal training for our other favorite senses—like art class for seeing or music classes for hearing—we don't get too much formal smell training. Scientists have found that with the right incentives (like electric shocks!), we can distinguish remarkably similar chemicals, even ones that look almost like mirror images of each other. You may have even run a similar experiment, after a long night of drinking a particular type of alcohol with unpleasant consequences, and afterward finding yourself incredibly sensitive to that particular smell.*

Some of the best and many of the worst smells we encounter in life are actually made by microbes. For example, the smell of the soil after the rain? That's geosmin, made by species of *Streptomyces*. And the smell of baking bread? That comes from the mixture of chemicals that yeasts make during the fermentation process.

The amount of the scent is important, too. A little bit of microbially produced isovaleric acid can smell cheesy, but too much and you might begin to realize it's the same chemical produced by microbes on sweaty socks. Even body odor, which you might think of as coming from your own cells, actually comes from microbes. We're actually secreting odorless chemicals from our armpits, and it just so happens that *Staphylococcus hominis* turns this into a chemical with a BO smell. Human (and human microbe) smells aren't all bad, though. There are some hypotheses that there may be some evolutionary advantage to identifying a partner who smells healthy and distinct from you.

Sometimes our relationship with our smells can get a bit complex. Trimethylamine, also known as TMA, gives off a strong fishlike odor. We don't produce TMA, but our gut microbes do if they are given the right precursors in our diet (especially choline, which you can find in red meat, eggs, soybeans, and a variety of other foods). The reason you don't notice most people smelling fishy, though, is because most of us have enzymes that convert the TMA to a less smelly form. However, there are a few people who lack the ability to make that particular enzyme and thus accumulate TMA in a disorder known as trimethylaminuria (fish odor syndrome). Examples of this disorder are also found in classic texts, including descriptions of Caliban in Shakespeare's *The Tempest*, and Satyavati, a character in the Indian epic the *Mahabharata*, ascribed to the legendary author Vyasa.

Satyavati's story is one of the best in all classic smell-related literature, in my opinion, so I'll relate it here, with some embellishments. The daughter of a fisherman, Satyavati helped ferry people across the river. Though she was kind and caring, people treated her cruelly due to her smell, and she resigned herself to a certain amount of loneliness, paddling people back and forth but never getting close to anyone. But then one day, a sage traveled across the river and was overcome with desire for her, and said to her, "Damn girl, you're hot. Let's bang."

A bit thrown off by the sage's sudden raunchiness and not entirely convinced by his overtures, she said, "You're a learned sage and I'm merely a girl who reeks of fish, so I think you'd probably not enjoy it."

And while the sage may not have learned whatever he was supposed to learn about chastity, he did seem well versed in persistence. "No problem!" he said, and just like that, he transformed her fishy scent to a sexy, musk scent.

Satyavati nearly dropped her oar. "Could this be . . . permanent?" she asked. The sage said, "Yeah, if you want. Now you ready to get it on?"

Satyavati looked around. "These reeds are not that tall, my dad is right over there—" and instantly he made a great fog appear. She gasped. He leaned in to kiss her. Her lips met his, almost. Except instead, she blurted out, "But you're just going to leave me! And in this society people will shame me for not being a virgin!"

The sage nodded. "Society is bullshit like that. I will use my sage powers to restore your virginity afterward." All of her conditions being met, they go for it. She's just so thrilled about the sex that she instantly has a child who immediately grows up, gives her a hug, and decides to go off to live in the forest. (And in yet another twist, that child is actually Vyasa, the legendary poet and author of the great epic itself!)

There're a few important lessons from this myth that help us understand sex, smells, and microbes. First, if a person is hitting on you, ask if they have magic powers. Second, smell alone won't necessarily dissuade someone from hitting on you. Some people actually lack the ability to smell TMA or lack the ability to smell at all. Third, be nice to people! Trimethylaminuria is a real (though rare) condition. Real people with this condition commonly endure rejection and social anxiety. Don't treat someone badly just because they have a smell you don't like.

But this is Nerd Nite so let's get to brass tacks—are microbes controlling your sex life? Probably not. But I hope this has given you a greater appreciation for all the scents in our lives, from the fragrant to the pungent, and encourages you to stop and smell the roses!

* *Editors' note: As undergrads, Chris threw a party and, among several unsavory concoctions, served apple pie shots: vodka and apple cider topped with whipped cream and cinnamon. After too many of them (which, honestly, is one), Matt puked*

all night and hasn't been able to even smell cinnamon since then without feeling nauseous. That was 28 years ago!

References

Li, W., et al. "Aversive Learning Enhances Perceptual and Cortical Discrimination of Indiscriminable Odor Cues." *Science* 319, no. 5871 (March 28, 2008).

McGann, John. "Poor Human Olfaction Is a 19th-Century Myth." *Science* 356, no. 6338 (May 12, 2017).

Rudden, M., et al. "The Molecular Basis of Thioalcohol Production in Human Body Odour." *Scientific Reports* 10, no. 12500 (July 27, 2020).

Jenny Bratburd is a science policy enthusiast with a PhD in Microbiology from the University of Wisconsin–Madison and currently works as an outreach coordinator for NASA's Health and Air Quality Applied Sciences Team.

Health and (Un)Wellness

Let's face it, the practice of medicine is a mess. I just spent $150 to take a class on how to use CPAP, which I already know how to use, because my insurance company is now monitoring the extent to which I use it. I didn't receive the machine my doctor prescribed to me so I'm still snoring at my misophonic wife (see the Mmm . . . Brains chapter). And I just learned on the internet that maybe I just need to tape my mouth shut and breathe through my nose. At the same time, there are lots of things to be excited about in our understanding of human health and medicine. In large part because of our improved ability to sequence DNA, we're starting to uncover so much about the role that microscopic, well, microbes play in our everyday life. For the same reason, we are at last making real progress in understanding the relationships between genetic variation and differences between people. It turns out that in all these topics, simple answers are generally wrong, so let's dig into the seedy underbelly and see what we can learn. Also, some of you might just want to know the cure for the hangover, so as always, we at Nerd Nite are here for you.

—Chris

NERD NITE NYC

MAGGOT THERAPY; or, How I Learned to Stop Worrying and Love the Bugs

by Avir Mitra, MD

Take two maggots and call me in the morning.

Diana Dupuy was an otherwise healthy woman who had bunion surgery. When her cast came off, she found that she had a wound that wasn't healing very well. Wounds need good blood flow to heal because blood brings oxygen, nutrients, and immune cells to the area. But in her case, the blood flow just wasn't that great all the way down at her foot. And so the tissue, instead of scabbing and healing, was starting to die. "I was like, oh my goodness!" recalls Diana. "I knew that wasn't good."

To treat this, a doctor needs to use a scalpel to debride, or cut away, dead tissue. Antibiotics need to be given, and the wound needs to be carefully bandaged. Her podiatrist did all of that, but the wound just kept getting worse. Diana recalls, "Weeks had gone by and the wound looked worse and worse. It looked nasty." It was starting to become black—the telltale sign of dying, necrotic tissue. The problem with dead, bloodless tissue is that bacteria can run amok, infecting the wound and using it as a launching pad to infect the rest of you. So, to her shock, her podiatrist told her that if things didn't turn around, they may have to amputate the foot to save her life. "I'm terrified at this point. I knew it was urgent, it was critical," states Diana. As you can imagine, she quickly became open to alternative ideas no matter how seemingly insane.

And that's when she stumbled across Ron Sherman, the maggot doctor. Ron is a real physician, but also believes in the healing power of worms.

Maggots are the larval form of the *Diptera* species—in other words, baby flies. It's not all that uncommon for maggots to infest and grow in poorly healing human tissue—it's called myiasis. Junior doctors across the world have spent countless, nauseating hours removing such maggots from rotting flesh wounds. But when Ron was a medical intern in the 1980s, he used what little free time he had wandering the wards, collecting unwanted maggots from patients' wounds, and studying them under the microscope. He started to read old texts and realized that people had actually studied maggots before. A surgeon in Napoleon's army noted that soldiers with maggots in their wounds seemed to fare better than those without. Similar doctor notes surfaced during the Civil War.

So Ron decided to conduct a study. He found 103 people with poorly healing wounds (despite maximum medical therapy) and somehow managed to convince half of them to let him put maggots into their wounds. What he found surprised everybody. In the patients who combined maggots with standard medical therapies, 80 percent achieved complete debridement compared with only 48 percent in those who politely declined the worms. "The results were so impressive that I continued working in that area," says Ron.

In 2004, after more successful studies, Ron applied to the FDA to get maggots approved for wound therapy. In 2007, maggots became the first living creatures to gain that approval. The FDA classified them as a "medical device," for lack of a better category. Ron then opened a business with his wife where the couple grows, cultivates, and ships sterile maggots to doctors like Diana's.

"I didn't know if it was going to make me vomit or pass out," remembers Diana. The maggots arrived at her doctor's office impregnated in a gauze pad. The doctor simply applied the pad to the wound and covered it with a breathable dressing. "The maggots were hungry and they immediately started doing their work. I panicked, I was like, Oh my God, look at all these maggots!" says Diana. "But it seemed like when I calmed down, they calmed down."

To her surprise, she didn't mind the maggots wiggling around in her foot. After just a few sessions, her foot appeared markedly better, and ultimately it completely healed. "I felt bad that I had to eventually kill them. They're doing this great treatment to me, and the reward is death," she said, laughing.

How is it that these brainless, disgusting maggots are doing a better job than doctors? Well, they're hungry.

Maggots are just baby flies that need lots of food, and it so happens that their favorite food is dead tissue. The problem with dead tissue is that it's pretty hard, like beef jerky, so maggots have evolved little prickly spines along their bodies, and when they crawl around a wound, they gently loosen the dead flesh and separate it away from the living flesh. It's actually a lot gentler and more effective than a surgeon's scalpel. Like our babies, maggots can't chew, so they vomit digestive enzymes onto the wound, dissolving the now loosened tissue into a tasty slurry, which they can slurp up. Amazingly, these processes spare the living tissue and only really affect dead tissue.

Turns out, maggots don't want bacteria in their food any more than we do.

They have evolved an enzyme called lucifensin, which sits in their gut and kills any bacteria they eat. This enzyme protects the maggot from infections, protecting us in the process. And if that wasn't enough, maggots stimulate the body to grow more blood vessels at the wound site, thereby allowing more crucial blood flow to reach the wound. How they do this is still a complete mystery. Thus, maggots are debriding wounds better than surgeons, killing bacteria better than some of our best antibiotics, and stimulating blood flow in wizard-like ways. And unlike most doctors, these maggots are dirt-cheap and willing to die for your wound.

Despite this, most patients and doctors remain hesitant to utilize the healing powers of maggots. According to Ron, "Less than 5 percent of patients who are destined for amputation are given a trial of maggot therapy, even though published studies show that 50 to 70 percent of those amputations could probably be prevented." Maggots are, after all, completely disgusting to most of us, yet they might just be what the doctor ordered.

Avir Mitra is an ER doctor and assistant professor of emergency medicine and education at the Icahn School of Medicine at Mount Sinai. He moonlights as a contributing editor at Radiolab *and has also written and produced stories for* Vice, The Pulse, *and* Here and Now.

NERD NITE NYC

HAPPY 13TH BIRTHDAY, NERD NITE! NOW GET YOURSELF TO AN ADOLESCENT MEDICINE DOCTOR

by **Dr. Nancy Dodson**

Editors' note: This presentation was given on March 9, 2019, to celebrate Nerd Nite NYC's 13th birthday. Perhaps Matt and Chris should have asked Nancy to update it for this book for Nerd Nite's 21st birthday, but alas, things that happen to a 13-year-old's body are much weirder than being able to be legally hung over.

When I decided to become a pediatrician, I had a Norman Rockwell painting in my mind. The painting, called *Doctor and the Doll*, depicts a little girl holding up a doll to her doctor as he obediently listens to the doll's chest with his stethoscope. I wanted to be the doctor in this Norman Rockwell painting.

Then I got to my pediatrics residency, and I realized something. The little girl in that Norman Rockwell painting would eventually grow up and become a teenager. She might try marijuana, or want to know about birth control, or have mental health struggles. And suddenly I found her much more interesting.

So I decided to specialize in adolescent medicine, a field within pediatrics that is small but full of mighty people. And this field has brought so many unexpected surprises to my life. Teens, it turns out, are endlessly interesting and challenging. And their needs change as the times around them change.

One of my job descriptions is to be a "period detective." Whether girls are getting their period too often or not often enough, whether it's too light or too

heavy, I am on the case. The most common period-related complaint I see is the complete disappearance of a period—the medical term for this is *amenorrhea*. Among my patients, this is often due to some intense stressor. The most common stressor is dieting and weight control, so I spend a lot of time trying to undo the dangerous messages they hear from social media, health teachers, and even well-meaning pediatricians who instill a fear of weight gain in a growing teen. But even psychological stress can cause periods to go awry. In the early days of Radcliffe College, it was noted that many girls got to college and developed amenorrhea. Now we know why—they were under intense psychological stress at a demanding college. But back then it was thought that their brains were working so hard that they were sucking the blood away from their uteruses!

One of the wonderful, unexpected blessings this field has given me is my work with transgender teens. I never expected to be a "gender doctor," but these teens have come into my life and I absolutely love helping them and their families figure out the right path forward. Many gender-diverse teens need only to explore gender expression with haircut and clothing. Others, who have true, deep, long-term distress from gender dysphoria, require hormonal treatment to transition genders.

As gender-diverse teens become a political punching bag nationally, I wish lawmakers could meet the patients I take care of. One transgender boy said to me, "I've been on testosterone for six months and three weeks now. And I used to be sad all the time, but now I'm not sad anymore. Because I'm finally getting to be the person I've always been inside."

So, in summary, here is what I want you to take away:

> Adolescent medicine is a wonderful field of medicine.
>
> Periods can go away when teens are stressed (which happens) or dieting (which happens but shouldn't).
>
> If you care about teens, don't ever vote for a Republican.

Dr. Nancy Dodson is an adolescent medicine doctor in Westchester County, New York. She specializes in eating disorder management and gender medicine, but sees a range of adolescent health issues. She doesn't have teen children, but she will very soon!

DETOUR
EXIT
EXIT

NERD NITE PHILLY

WHAT YOUR DNA SAYS ABOUT YOU

by **Shweta Ramdas**

Have you noticed how all of a sudden the words (well, acronym and word, respectively) *DNA* and *genes* seem to be everywhere? My mum would earlier use "It's in your blood" to explain away any of my flaws. But these days she says, "You have my genes." It wouldn't be hyperbole to say that genetics is revolutionizing what we know about ourselves. This is largely because it has become really cheap to sequence DNA—just sample any living thing around you, extract its DNA, dab a few drops into a machine, and voilà. You have millions of letters of the organism's genome sequence before your eyes. And these days DNA sequencers are small enough to be mistaken for a USB drive.

But how exactly is this "revolution" happening and why should we care? Simply put, genetics tells us a lot about the history of our species; it allows us to learn things about those of us who are alive today (whether we want to or not); and it can tell us things about the individual *you*.

When our ancestors (*Homo sapiens* senior) left their family homes in Africa to see what the rest of the world looked like, they ran into their Neanderthal friends in Europe. The word *Neanderthal* is used as an insult—these were an ancient human-like species that lived in Europe 50,000 years ago that we, the superior *Homo* of the *sapiens* variety, outlived because of our sophistication, intelligence, and general awesomeness. Clearly. But genetics has unfortunately poured cold water on this very alluring story.

We've been able to look at Neanderthal DNA (a discovery that won the

Nobel Prize in 2022!), and when we compare it with ours, we see that our genomes are filled with chunks of Neanderthal DNA. It's like reading *Harry Potter* and finding passages about Sauron and the rings of power in there . . . There is only one explanation for this—our ancestors' European backpacking trip involved some hanky-panky, and those pesky Neanderthals weren't as unsophisticated (or unattractive!) as we assumed. All non-African populations carry 2 percent Neanderthal DNA in our genomes. Even my dad discovered through his genetic test that he carries 1.8 percent Neanderthal DNA, which is less than 96 percent of all tested people. I guess my mum was wrong about my dad!

What most of us are interested in when we think of genetics, though, is how it affects our health. We all want to be able to blame our genetics (code for "our parents") for our weight and most other sources of our vanity. Biology is rarely as cooperative, though. Of the most common traits (and diseases) we think about—let's say heart disease, body weight, and ~~the president you voted for~~—there is no single gene that determines the trait. In other words, there's no magic predictor that will tell you if you will get a heart attack or not. This is because most common diseases are *complex*, as your risk is determined by an interplay between your genetic risk factors and your environment (lifestyle, diet, culture). Moreover, there's no single gene for any of those traits. For example, there are 12,000 parts of your genome that can affect your height alone!

As much as it pains me as a geneticist to say this—your genetics don't always tell you very much. Pick any common human trait that interests you; for me it's body weight. Any single genetic factor would increase your weight by very little (one gram, let's say). Most of us have parts of our genome that tend to increase our weight and other parts that tend to decrease it, and thus our overall genetics predisposes us to a little more or a little less than average weight. However, about 10 to 20 percent of people may possess mostly trait-increasing genes, and for them, genetics would make them very likely to be overweight. A great example of something like this is seven-foot-six-inch-tall basketball player Shawn Bradley. An intrepid geneticist found himself next to Bradley on a flight (any guesses on what the airline with the presumably great legroom might have been?) and asked if he could sample his DNA. Turns out Bradley has a huge number of genetic factors that each increase his height by just a little amount, giving him a big genetic advantage in the height (and the basketball) department. The chances of a person having so many height-increasing variants is less than one in a million. And the chances of a geneticist trying to weasel your DNA out of you are much higher.

While we tend to think of genes as our destiny, a more accurate way of thinking about genes is as a nudge down a certain path. You can still blame your parents for your flaws, though; just leave the genetics out of it.

If I have succeeded, you have been mesmerized by the sheer coolness of anything genetics and can't wait to spit into a tube and get a company to sequence your DNA for you. These tests can be fun, but do take those companies that promise you a healthy "genetics-based exercise routine" with a big bucketful of salt. Genetics is everywhere, and capitalism is eager to cash in.

Shweta Ramdas is a geneticist working at a liberal arts university in Bengaluru, India. Her day job involves staring at DNA sequences on her computer and convincing her unsuspecting students of the wonders of genetics.

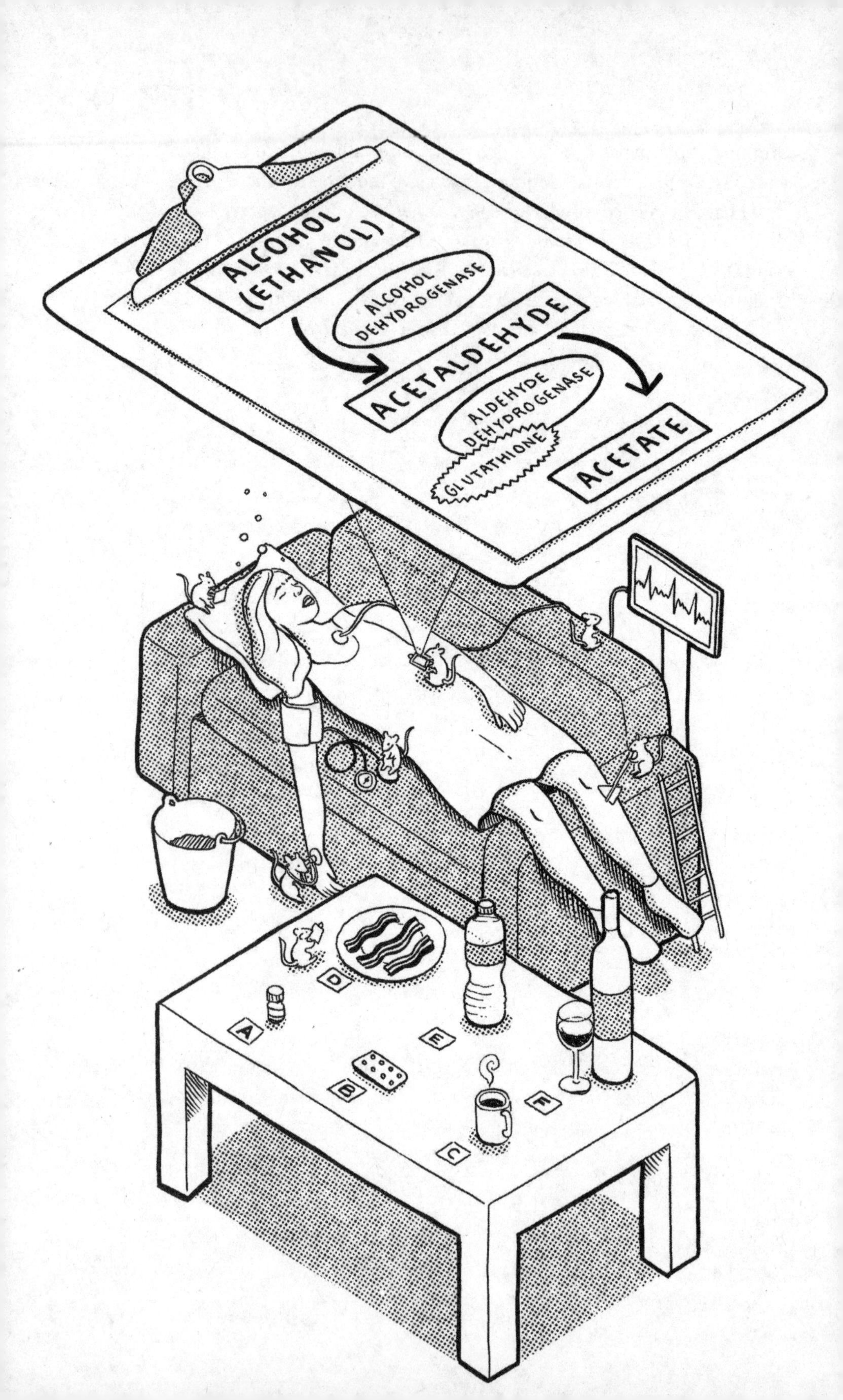
ALCOHOL (ETHANOL)
ALCOHOL DEHYDROGENASE
ACETALDEHYDE
ALDEHYDE DEHYDROGENASE
GLUTATHIONE
ACETATE
A
B
C
D
E
F

NERD NITE NYC

THE SCIENCE OF THE HANGOVER

by **Paula Croxson, DPhil**

The alcohol hangover: a "feeling of general misery."[1]

Is any of this experience familiar to you? Unconsciousness, inability to move—other than your eyes—thirst, headache, nausea, dizziness, cognitive impairment . . . and guilt. If it is, you have probably overindulged in alcohol consumption at some point in your life. And if you haven't—well, then I hate you.

My name is Paula Croxson and I am famous for having the worst hangovers in the world. My hangovers are so bad that they are named after me; they are called Croxson specials. Trust me, I'm a doctor. Well, a PhD, but close enough.

This affliction has led me, over the course of my adult life, to go on a journey of rigorous scientific and self-discovery, in search of the causes of—and possible cures for—the hangover.

So what is a hangover? Their scientific name is veisalgia: the unpleasant next-day state following an evening of excessive alcohol consumption. Veisalgia leads to an average of 0.6 to 0.7 workdays spent drunk or hung over per worker each year[2] in the UK, where we're not unfamiliar with heavy drinking. Honestly, that . . . doesn't seem that much to me? But when you add up the cost of all those lost partial days, it could have a pretty big economic impact.

1 Joris Verster, "The Alcohol Hangover: A Puzzling Phenomenon," *Alcohol and Alcoholism* 43, no. 2 (January 8, 2008): 124–26, https//doi.org/10.1093/alcalc/agm163.

2 https://www.ias.org.uk/2019/06/25/hangovers-are-a-financial-headache-for-the-uk-economy/.

To be honest, there isn't much research out there on hangovers at all, largely because they're considered to be our own fault. But there are some major reasons, aside from the economic ones, to research hangovers. A high incidence of hangovers is associated with an increased risk of dying of heart disease[3] or developing alcohol use disorder.[4] So people are still studying hangovers, even though, well—let's just say that I don't get a lot of sympathy for mine.

To start my research, I looked into what makes a hangover, a hangover. A study carried out in 2007 on two groups of people, US urban college students and professional mariners in a Swedish academy, asked about the most common symptoms of hangover in a scholarly effort to develop a new hangover scale.[5] The participants said that the number one most common symptom of hangovers was "the feeling of hangover." Great. So insightful.

But the study participants also mentioned many other common hangover symptoms such as thirst, headache, tiredness, dizziness, loss of appetite, stomachache, nausea, racing heart. So I, being the rigorous scientist that I am, decided to compare the results of this study to my own symptoms of the Croxson special. And my symptoms are this: vomiting, vomiting, vomiting, vomiting, every 20 minutes from 7 AM until 7 PM until I pass out from exhaustion.

I think you will agree that my hangovers are worthy of deeper, more intensive research than other people's hangovers, because I am truly suffering.

So then I started to look into whether there was a *cure* for the hangover . . .

. . . and it turns out that there *is* a cure for the hangover. And the cure is . . . not to drink alcohol at all. Which is *ridiculous*. So let's just exclude that immediately.

If there isn't another cure, are there any effective *treatments*? In another survey, where they asked Dutch students—and they should know, right?—what the most effective treatments were for them, the students mentioned things like drinking water, sleeping a lot, having a nice greasy breakfast, painkillers, and tea or coffee. Of course, these are all anecdotal, so I looked into all of these potential treatments and discovered that *only one* has a scientific basis. And

3 J. Kauhanen et al., "Frequent Hangovers and Cardiovascular Mortality in Middle-Aged Men," *Epidemiology* 8, no. 3 (May 1997): 310–14, https//doi.org/10.1097/00001648-199705000-00014.

4 Damaris J. Rohsenow et al., "Hangover Sensitivity After Controlled Alcohol Administration as Predictor of Post-College Drinking," *Journal of Abnormal Psychology* 121, no. 1 (2012): 270–75, https//doi.org/10.1037/a0024706.

5 Damaris J. Rohsenow et al., "The Acute Hangover Scale: A New Measure of Immediate Hangover Symptoms," *Addictive Behaviors* 32, no. 6 (June 2007): 1314–20, https//doi.org/10.1016/j.addbeh.2006.10.001.

that is: drinking more alcohol. That's right! The-hair-of-the-dog-that-bit-you is actually proven to be scientifically justified.

Here's why: When you drink a lot of alcohol, you may also end up with elevated levels of methanol in your body because it relies on the same pathway as ethanol (the main ingredient in our alcohol drinks) to be broken down, and that pathway is overwhelmed after a big night of drinking. It also shows up as an impurity in some alcoholic drinks. Methanol is very toxic and likely leads to some of the symptoms of hangover.[6] If you are ever unlucky enough to be hospitalized for methanol poisoning, they will treat you by giving you . . . ethanol. That will bind to the methanol and neutralize some of its effects.[7] So self-medicating with more alcohol does work, but unfortunately people who rely on the-hair-of-the-dog are more likely to binge-drink, more likely to drink more frequently, and theoretically might be more likely to develop alcohol dependence.[8] So I couldn't recommend that.

If the only treatment is too risky to contemplate, what about hangover *prevention*? It's a well-known fact that we should be drinking water before, during, after—and perhaps *instead* of—drinking alcohol. There is a scientific basis for drinking more water when you are drinking alcohol, and it's because when you drink alcohol you decrease your levels of vasopressin, a chemical in your kidneys that allows you to retain water.[9] When your vasopressin levels are low, you pee more, and, well, I'm sure you've heard about "breaking the seal."

Unfortunately, when I looked into that further, it turns out that drinking a lot of water doesn't really do anything to alleviate the symptoms of a hangover. That headache, for example, isn't caused by dehydration.[10]

The one thing that drinking water might help with is motor impairment.

---

6 A. W. Jones, "Elimination Half-Life of Methanol During Hangover," *Basic and Clinical Pharmacology and Toxicology* 60, no. 3 (March 1987): 161–238, https://doi.org/10.1111/j.1600-0773.1987.tb01737.x.

7 Kurt Anseeuw, Marc B. Sabbe, and Annemie Legrand, "Methanol Poisoning: The Duality Between 'Fast and Cheap' and 'Slow and Expensive,'" *European Journal of Emergency Medicine* 15, no. 2 (April 2008): 107–09, https://doi.org/10.1097/MEJ.0b013e3282f3c13b.

8 T. B. Baker et al., "Addiction Motivation Reformulated: An Affective Processing Model of Negative Reinforcement," *Psychological Reviews* 111, no. 1 (2004): 33–51, https://doi.org/10.1037/0033-295X.111.1.33.

9 G. Eisenhow and R. H. Johnson, "Effect of Ethanol Ingestion on Plasma Vasopressin and Water Balance in Humans," *American Journal of Physiology* 242, no. 5 (May 1, 1982): R522–R527, https://doi.org/10.1152/ajpregu.1982.242.5.R522.

10 J. C. Verster et al., "Alcohol Hangover Amongst Canadian University Students: Can Hangover Immunity Be Really Claimed?" *European Neuropsycholpharmacology* 25, suppl. 2 (September 2015), S603, https://doi.org/10.1016/S0924-977X(15)30850-6.

Have you ever had shaky legs the next day? That's when I know it's been a really big night. One really cute study in which they painted the feet of mice and tested their ability to walk in a straight line after sobering up from alcohol consumption showed that giving the mice water reduced their motor impairment.[11] So that's encouraging, but it's not going to cure the Croxson special.

Another theory of hangovers is that they are caused by the impurities in the alcohol you drink. The amount of congeners—organic compounds that are found in alcohol drinks and give them their taste—varies quite widely with the type of drink. The congener hangover scale was developed after research indicated that the higher the levels of congeners, for example in red wine or whiskey, the worse the hangover.[12] However, when I lined up my personal scale with the congener hangover scale, I found very little overlap, so I guess I will have to continue this line of personal research further . . .

At this point I had reached a dead end in my research, and I was really out of options. Until one day, in the throes of an unusually bad Croxson special, in desperation I took one of my seasickness tablets. And . . . it kind of worked a little bit. And when I read the label on the package, it turns out that my seasickness tablets are actually antihistamines.

Why is that important? Well, let's look at the chemical breakdown process that happens when we consume alcohol.[13] In simplified terms, there are two major breakdown steps for alcohol: First it gets broken down by the enzyme alcohol dehydrogenase into acetaldehyde; then it's broken down further by a second enzyme, aldehyde dehydrogenase, into acetate. That second breakdown step is very important, because acetaldehyde is highly toxic and a known carcinogen.

Some people don't manufacture very much of that second enzyme, aldehyde dehydrogenase. These are the people who turn bright red when they drink alcohol, and experience other unpleasant things like nausea, itchiness, and puffiness. This is sometimes called alcohol flush reaction, and it's really uncomfortable and unpleasant and has been linked to an increased risk of cancer,[14] and also a decreased likelihood of people wanting to have their photo taken.

---

11 Analía G. Karadayian and Rodolfo A. Cutrera, "Alcohol Hangover: Type and Time-Extension of Motor Function Impairments," *Behavioural Brain Research* 247 (June 2013): 165–73, https://doi.org/10.1016/j.bbr.2013.03.037.

12 Verster, "Alcohol Hangover."

13 "Alcohol Metabolism: An Update," *Alcohol Alert* (National Institute on Alcohol Abuse and Alcoholism) 72 (July 2007), https://pubs.niaaa.nih.gov/publications/aa72/aa72.htm.

14 https://www.niaaa.nih.gov/publications/alcohol-flush-reaction.

That reaction? It's a histamine response. And so, I wondered, is the alcohol flush reaction something that might be happening for me? Well, no—I don't go red when I drink. So it's not that I'm not making enough aldehyde dehydrogenase. But then I found out that there is a thing called a cofactor, which makes an enzyme work better, and there is a particular cofactor called glutathione, which you need in order for your aldehyde dehydrogenase to work.[15] It turns out that alcohol consumption leads to a depletion of glutathione, and that leads to a buildup of acetaldehyde.

And so, I reasoned, it's possible that I am having a similar experience to the alcohol flush reaction, just delayed until the next morning because it's happening more gradually for me. We might call it the Croxson flush reaction. And if that's the case, is there something I can do about it?

And that brings me to the *really* cool part . . . you can actually buy the precursor to glutathione online. It's called N-acetyl cysteine (or NAC), and you can buy it wherever you like to get your supplements in bottles of 200 tablets, and once you take one it'll be metabolized into glutathione.

So I ordered it, and next time I indulged in a few drinks too many, I made sure I took an NAC tablet before I went to sleep. And *my hangovers went away.* I did get a bit more of a headache, because, as I mentioned earlier, that headache isn't caused by dehydration. It's likely caused by the final breakdown product of alcohol: acetate.[16] But I was able to get out of bed and go about my day, and that was *amazing.*

I looked into the use of NAC extensively and I found only one clinical trial,[17] which concluded that it wasn't effective in reducing hangover symptoms. I will say that they used a 52-point scale to measure hangovers, where 1 is the least bad, and 52 is the worst, and most of the participants in the study had a median of 13-point hangovers. For comparison, when I took the scale, I got 52. So perhaps they need to find people with worse hangovers to test it on. There were also a few "mild" side effects, including gastrointestinal side effects and headaches, that some of my friends who tried NAC have also experienced.

15 Barbara L. Vogt and John P. Richie Jr., "Glutathione Depletion and Recovery After Acute Ethanol Administration in the Aging Mouse," *Biochemical Pharmacology* 73 no. 10 (May 15, 2007): 1613–21, https://doi.org/10.1016/j.bcp.2007.01.033.

16 Christina R. Maxwell et al., "Acetate Causes Alcohol Hangover Headache in Rats," *PLoS One* 5, no. 12 (December 31, 2010): e15963, https://doi.org/10.1371/journal.pone.0015963.

17 Veronica Coppersmith et al., "The Use of N-acetylcysteine in the Prevention of Hangover: A Randomized Trial," *Scientific Reports* 11, no. 13397 (June 28, 2021): https://www.nature.com/articles/s41598-021-92676-0.

Some of those friends are not very happy with me for recommending NAC so enthusiastically.

So I'm not actually recommending that you try this hangover treatment because it's scientifically ineffective and might lead to some unpleasant side effects, but maybe, just maybe, I have at least found the cure for the Croxson special.

Paula Croxson, DPhil, is a neuroscientist, science communicator, and storyteller. She is also an open-water swimmer and the flautist in two rock bands. The swimming is apparently for "fun."

NERD NITE DC

YOU AND YOUR MICROBIOME: Say Hello to Your Little Friends

by Rebecca B. Blank, MD, PhD

You are not alone. In this increasingly disconnected world filled with virtual communities, you can rest assured there is at least one community that has your back: your microbiome, those trillions of microorganisms that live within and upon you.

Indeed, there are approximately 100 trillion microorganisms (mainly bacteria), representing as many as 30,000 different species, living in every crevice, nook, and mucosal cranny of your body that you can imagine. Have you washed your belly button lately? Didn't think so. These trillions of microorganisms that live both on and within you are, for the most part, your friends. One could even say they're your awesome roommates who take care of your plants when you forget, while being provided, in return, with a calm and stable living environment where the rent is paid on time and food is always in the fridge. In the worst-case scenarios, your microbiome, if not treated with respect or for other reasons not quite understood, like any somewhat unstable roommate, can wreak havoc and cause disequilibrium to your digestion, your immune system, and even your brain chemistry.

By now, most people have heard at least something about the microbiome, and in particular the microbiota that live in your gut. (If not, check out any Activia yogurt advertisement.) The gut microbiome is the most fashionable organ of the moment! Because these communities in the gut, on the skin, and on every other mucosal surface of your body comprise trillions of cells, your microbiome actually outnumbers the total number of cells in your body by about ten to one. And because these microbial communities work together to maintain the general health of their human niches, they have been described as an important additional organ, one you didn't learn about in high school biology class, but one that you really need if you hope to maintain a state of general well-being.

At first thought, you might wonder why you should be thankful that microbes line your gut and skin. Aren't bacteria *bad*? Don't they cause disease? Well, yes and no. Some do cause disease, but the bacteria that line your mucosal surfaces are so well suited to your personal environment that they outcompete potential pathogens for resources and mucosal real estate. And they also induce a homeostatic immune response that allows your body to help fend off those pathogens that actually do breach the physical barrier of your skin or gut lining. The human microbiome and, for this article's focus, the gut microbiome, have been associated with several healthy outcomes. Gut microbes are essential in breaking down a number of foodstuffs that humans would otherwise not be able to digest. Studies have shown that the nutritional value of the food we eat is, in part, influenced by the structure and function of our gut communities. Not only do gut microbes synthesize several vitamins we cannot produce on our own, they can also assist in our absorption of key minerals such as iron. Importantly, they also can determine how many calories we obtain from every low-fat Special K cereal bar we consume. In fact, researchers have shown that the microbial communities of obese and lean people differ substantially in terms of the number and type of different species and in the microbes' expression of genes involved in metabolism. While obese people seem to harbor a less diverse population of microbes, they do tend to harbor greater concentrations of microbes that break down food really well, extracting every last calorie from that low-fat snack bar, whereas the gut microbes in lean people tend not to be quite so efficient.

It's a horrifying concept when you think about it: The microbes of obese people are actually helping them fatten up! And those damn skinny people can eat all they want and not gain weight because they have inefficient microbes! Well, luckily it appears that this is not altogether the case. Not only does outward physical appearance, lean and not-so-lean, predict the composition of gut communities, but so does diet. A high-fat, high-calorie diet tends to drive the expansion of very efficient "obese" microbes, whereas a low-fat diet drives the expansion of "lean" microbes. We may actually have the ability to modify our gut microbiota to some extent by changing our diet alone. Studies in mice whose guts have been transplanted with human fecal microbial communities have demonstrated that the composition of the species in mice transplanted with "obese" microbiota can shift to a "leaner" microbial composition if they are fed a lean diet. And the same is true for mice with "lean" microbiota that are fed an obese diet.

Perturbations in the gut microbial community have been linked to a number of devastating conditions, including type 1 and type 2 diabetes, asthma, inflammatory bowel disease, intractable *Clostridium difficile* infection, inflammatory arthritis, and even some behavioral and psychiatric disorders. Microbes begin to colonize their human niches at birth and are pretty much settled by the time a child is three years old. Researchers have found that the mode of childbirth has a significant effect on the composition of the gut microbiomes of newborns; babies who are born via vaginal delivery tend to be colonized with microbes found in the birth canal and the mother's digestive tract, whereas C-section babies tend to be colonized with microbial communities found most prevalently on skin. Some researchers postulate that this perturbation in colonization of the newborn gut may have consequences for susceptibility to allergic responses and asthma later in life, although studies of this sort have had few subjects and are not yet definitive.

It is well understood that the gut microbiome influences the immune response during homeostasis and during infection. In fact, the normal gut flora are responsible for inducing human regulatory immune cells that limit the amount of tissue destruction during an active infection with a pathogen, be it bacterial, viral, or even a parasitic worm. When people take powerful antibiotics, their gut microbiome numbers and composition are drastically perturbed because antibiotics are indiscriminate and kill most microbes that come into contact with them. The problem is that when the healthy community of microbes gets perturbed, it is easier for a pathogenic strain to gain a foothold in the niche that was once occupied by the healthy strain. Multi-drug-resistant *Clostridium difficile* is one such bacterial species that is really shifty. It waits until the healthy microbes are weakened by a bout of antibiotics and then moves in, overgrowing and overstaying its welcome in your intestine. This usually happens in hospital settings where patients are subjected to multiple rounds of different antibiotics. Hospital-acquired *C. difficile* can be unbelievably hard to get rid of because it is often resistant to most antibiotics (hence its ability to take over and wreak havoc in the first place). Doctors found that one way to push out these obnoxious gut-community invaders was to give patients fecal transplants from a closely related family member in the hope that the transplanted healthy fecal community will be able to overpower and crowd out the offending pathogen. This is totally disgusting to think about but actually worked very well. Studies have found that while your own microbiome is unique to you, it most closely resembles members of your own household. I will let the reader cogitate on why household members share so many members of their gut microbiomes, but might I mention the sign in all restaurant bathrooms imploring one to wash one's hands?

(As an update to the original article, now companies have cashed in on—and the FDA has regulated—fecal transplant therapy. A *C. difficile* sufferer no longer has to rely on a willing family member but can get the fecal "gift" commercially.)

There is also compelling evidence linking the composition of the gut microbiome to behavioral and psychiatric states. It is well known that certain bacterial infections can cause psychiatric disorders. For example, infection with *Leptospira* can cause manic and psychotic symptoms that can only be treated with antibiotics that eliminate the infection. Neuropsychiatric disorders in children have also been linked to streptococcal infections, so consider yourself lucky if you had your tonsils taken out when you were younger. You missed out on all the opportunities to become OCD after a strep throat infection. It is no surprise, then, that the microbes that line your gut may also affect behavior. Scientists have shown that mice that are raised germ-free, meaning they do not have any microbiota, exhibit less anxiety than their germ-replete controls. Not as convincing, but nonetheless intriguing, are numerous reports linking diet to manifestations of psychiatric disorders such as schizophrenia, depression, attention deficit hyperactivity disorder, and even autism. By the transitive property, the composition of the gut microbiota is also influenced by diet and is very important in metabolizing various nutrients—hence the postulation of a link between the microbial communities in the gut and psychiatric disease.

Building upon the gut-brain connection, there are several reports that children who suffer from autism spectrum disorders also suffer from a variety of gastrointestinal disorders. There have been some studies in which children on the autism spectrum have a greater representation of the bacterial family Clostridiales than healthy controls. Even more compelling is evidence from a small study of autistic children showing that treatment with antibiotics actually improved autism symptoms. This evidence suggests that the microbes that may be responsible for the autistic symptoms are being eliminated by the antibiotic treatment and are no longer able to produce whatever neurotoxic metabolites were causing the children to experience those symptoms. Based on all the interesting connections that researchers are finding between the gut microbiome and various disease states, it's no wonder that the microbiome is such a hot topic or that researchers—and, let's face it, companies—are investing so much time and energy investigating ways to tweak the microbial communities that live within us to our advantage.

Some potential therapeutics already on the market are so-called pre- or probiotics. Probiotics are basically living microbes that are supposed to confer a health benefit if consumed in large enough quantities. Prebiotics are compounds that can be administered to a microbial community in order to induce the growth of a specific "healthy" subset of microbes. Both of these microbiome therapeutics are supposed to promote a general shifting of your microbial communities from a dysbiotic state (in which the microbiome is causing disease) to a healthy and happy state. Given the shifting of gut communities via changes in diet and administration of antibiotics, pre- and probiotics are gaining traction among consumers and scientists.

Further down the pipeline we can envision a therapy in which a person swallows a bolus of genetically engineered microbes that carry vaccine antigens or therapeutic drugs directly to the mucosal surface of interest. Also, we can envision a future whereby we take a drug that targets the microbes that live with us, as opposed to one that affects our human cells. In that sense, we are not just an organism but a superorganism, one made of many simpatico communities both highly evolved and single celled. So the next time you feel alone and disconnected, think about your microbiome and realize that there are countless communities very dear to you that are looking out for your best interests.

Rebecca Blank, MD, PhD, is an immunoparasitologist turned mucosal immunologist turned rheumatologist. She finds the interactions between the human immune response and the organisms that live within and upon us unbelievably fascinating. Due to her creepy-crawly background, she also likes grossing people out with her stories of brain-eating hot springs parasites and other esoteric diseases.

NERD NITE GREENVILLE, NC

PENIS OR VAGINA? 'TAIN'T THAT SIMPLE!

by Krista A. McCoy, PhD

What is sex? Well, it's fun! I know that much. It's also how we make babies and it's how we describe our body type—boy or girl. But what determines whether we are considered a boy or a girl? That is actually super simple! The first thing the doctor does when we are born is look directly at our genitalia and declare what sex we are. That defines us for life. They don't look at the chubby little face or skinny legs, or notice the lack of a tail. They don't declare, "It's a human." Instead, they categorize us into two groups: Penis and Vagina. You can't have both, right?

This is literally not brain surgery. Guys have penises, testicles, sperm, and XY chromosomes. That is, males carry heteromorphic (different-looking) chromosomes. Ladies have vaginas, ovaries, eggs, and two X chromosomes. Right? Well, yes, these traits are part of the biological definition of sex. But this isn't true for all species. In birds (and many other animals) the females are the heteromorphic sex (ZW) and males the homomorphic sex (ZZ). So not all males have two different-looking chromosomes that determine which reproductive organs develop. In platypuses, males carry five X and five Y chromosomes. See, it is all very straightforward.

For many species it is not as simple as having two similar or two (or 10) different sex chromosomes; some do not have different-looking chromosomes at all. For example, turtles don't have visibly different sex chromosomes; their sex is determined by their temperature during development. Temperature influences the hormones that are secreted and the hormones determine whether a generic baby gonad becomes testes or an ovary. Many fish species change sex naturally; if they didn't, their entire species would go extinct. Blue-headed wrasse (*Thalassoma bifasciatum*) hatch as females and live in a harem (many females, one male). When the dominant male dies, the largest female completely metamorphoses into a male. The yellow-and-brown-striped female becomes larger and its head turns blue while its body turns bright yellow. Ovaries that previously produced eggs turn into testicles that produce sperm. Sex-changing females initiate male sexual and aggressive displays within minutes of the removal of the dominant male. See, it is all very, very simple.

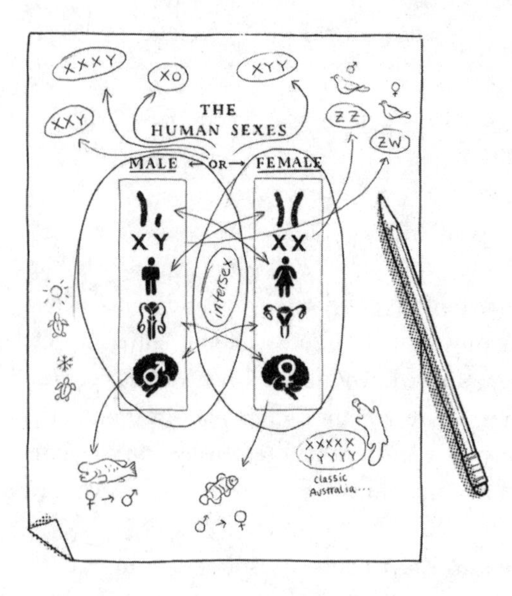

Even in humans, variation in chromosome numbers leads to variation in how our bodies develop. People can have multiple sex chromosomes (XXY, or XYY) or only one chromosome (XO). People with these different chromosome variants can have a variety of sexual traits that don't fit neatly into our two-category system, and although we can characterize external traits, we do not know how this variation in chromosome number affects other axes of sex like the sexual development of the brain.

Even when human chromosomes are shaped the way Congress and some governors prefer (either XX or XY), there are specific genes on those sex chromosomes, or others, that may not function in a typical way during development. This could be a result of changes in the DNA sequence or in the protein products they encode. Gene functions are modified by random mutation, by developmental exposure to certain pollutants, and naturally through mechanisms we don't yet know. Known mutations can cause androgen insensitivity—basically the protein that usually processes androgens (like testosterone) doesn't work as efficiently. Even though there is testosterone floating around the body in a male-typical way, the genitalia develops more like a female. Such a person would have XY chromosomes, but the doctor would have seen a Vagina. It is all very, very very simple.

Now in the case above, we have male sex chromosomes in a person with clearly female genitalia. As you might have surmised, it is also possible to wind up somewhere in between. Altered hormone synthesis during development can cause incomplete masculinization, which can result in ambiguous genitalia (the birthing doctor is confused about which sex to proclaim—because it turns out we are not always just a Penis or Vagina), and this is considered an intersex condition. There are numerous known causes of a diversity of intersex conditions, and roughly 2 percent of people (that is two in 100) are born with sexual characteristics that don't fit nicely into our two-group system. For example, we know that exposure to certain pollutants during development can cause penis deformities. Easy peasy.

Which brings us to the brain. While we have a general understanding of the genetic drivers of sexual differentiation of the external genitalia, we know almost nothing about if or how they relate to sexual differentiation of the brain. In fact, we don't know what hormonal cascades organize or activate sexual orientation, gender identity, or gender expression in the brain, but I think we should all be able to agree that our brains are not simply extensions of our penises and vaginas. Alas, because we must decide Penis or Vagina, children with some intersex conditions are commonly raised as girls. However, at the onset of puberty, they experience a large increase of testosterone typical of male puberty, which comes from their undescended testes. They develop typical male features, including increased body muscle mass, deepening of the voice, and enlargement of the external genitalia (which was considered the clitoris prior to puberty) into a functional penis; their testicles descend, and there is absence of breast development. I'm going to suggest that these kids don't need state officials making their personal or medical decisions for them. Just sayin'.

Once we know a little about the complexity of sexual development, the idea that someone's birth sex might not align with the way they feel seems less surprising. Indeed, many elected officials seem to spend a lot of time thinking about other people's genitalia and whether their behavior matches what might be in their pants. Categorizing everyone you see as a Penis or a Vagina might not really help anyone. It just 'tain't that simple.

Krista: wife, mother, daughter, scientist, antiracist, LGBTQ ally, het cis woman, opinionated, leader, happy.

NERD NITE NYC

THE MODERN STUDY OF GENETICS IS FULL OF TWISTS AND TURNS

by **Dr. C. Brandon Ogbunu**

Ordinary life is pretty complex stuff.

—Harvey Pekar

I'll come right out and say it: DNA is the coolest, most important, and most powerful piece of organized matter in the universe. It is, as we all know, the chemical that many call the recipe for constructing and operating many aspects of all living things on Earth.

From archaea that live in hot springs to the blue whales of the ocean, everything that we call a living thing contains DNA in almost every cell.

DNA does it all: It provides instructions for how to break down the sugars in our food, how an embryo develops, how neurons fire, how parasites cause illness.

I'm an evolutionary and population geneticist who studies how genes—and the proteins that they encode—evolve and change. That is, I'm interested in the story behind and consequences of the DNA that lives in our cells: how did it get there, how does it change, and how does this influence the great diversity that defines life on Earth.

But I have a confession to make. For all of its greatness, DNA is misunderstood. While we appreciate it for all its gifts, many of its powers have been oversold and misrepresented.

How did this happen? First, there are no singular villains in this story. Sure, the history of the field of genetics contains some misguided (or downright rotten) people who touted DNA as the answer to all of our questions about who we are, how the universe runs, how societies are built, why diseases happen, and why I'm so bad at fantasy football. But the problems with this history are not about any individual; rather, they are rooted in generations of mistakes and foibles (like most big problems, if you think about it). A full treatment of these problems is too long for any one article (or book), but for our purposes, we can begin with one of my big heroes, Gregor Mendel.

1. Mendel the Monk

Gregor Mendel conducted some of the most important experiments in the history of science. Mendel was an Austrian monk who, starting in the 1850s, performed a set of breeding experiments in pea plants that would give birth to the formal study of genetics. Through thousands of laborious, painstaking breeding tests, he measured and counted everything about these plants—their sex, the number of offspring, and, most important, the traits of parent pea plants and their offspring, including seed shape, flower color, pod shape, and others.

Through measuring everything carefully, Mendel was able to develop the closest thing to a set of laws—not entirely unlike Newton's laws of motion—that govern how heredity works, how genes and traits are passed down from generation to generation. These laws would transform genetics from a qualitative science where we all observe (using our senses) how traits are passed down to a truly quantitative, even somewhat predictive science. It is safe to say that Mendel's breakthroughs were critical for developing a basic understanding of how it all actually worked. I can't stress enough how elegant the experiments were and how important the laws that he developed were. Even further, Mendel's findings help us understand (and even treat) debilitating and harmful human diseases like sickle-cell anemia and Tay-Sachs disease.

So what is the problem?

The problem is that Mendel's experiments gave us all a false hope for how DNA is supposed to work in general. In some ways, we've all been chasing those Mendel experiments, or rather using those pea plant experiments as the basis of our expectations for how genes are supposed to work. As it turns out, the laws of Mendel don't apply quite so neatly to many of the traits that we care to measure in the biological world.

2. Genetics Is . . . Complicated

A major part of Mendel's genius was in his choice of system—he chose a plant (the pea plant) that features "true breeding," meaning the offspring's traits follow very directly from the genes that they inherit from their parents. Evolutionary geneticists might say that their "genotype-phenotype" map is linear—that is, the genetic makeup of the organism (pea plants in Mendel's case) maps onto observable traits (phenotype). In addition, many of the traits that Mendel studied were the product of a small number of genes located in a single location on the genome. More often, however, traits that we study are the product of many different genes (often hundreds!) working together.

Take human height. Height certainly appears to be related to what is in our genes: Tall people tend to have children who are tall. But some estimates of the genetic underpinnings of human height suggest that maybe thousands of variants of genes (alleles) might make a contribution. This means that human height is not "Mendelian," as in the product of a small number of genes, but rather a "complex" trait composed of many different genes.

HOW WE THOUGHT DNA CONTROLLED MOST TRAITS

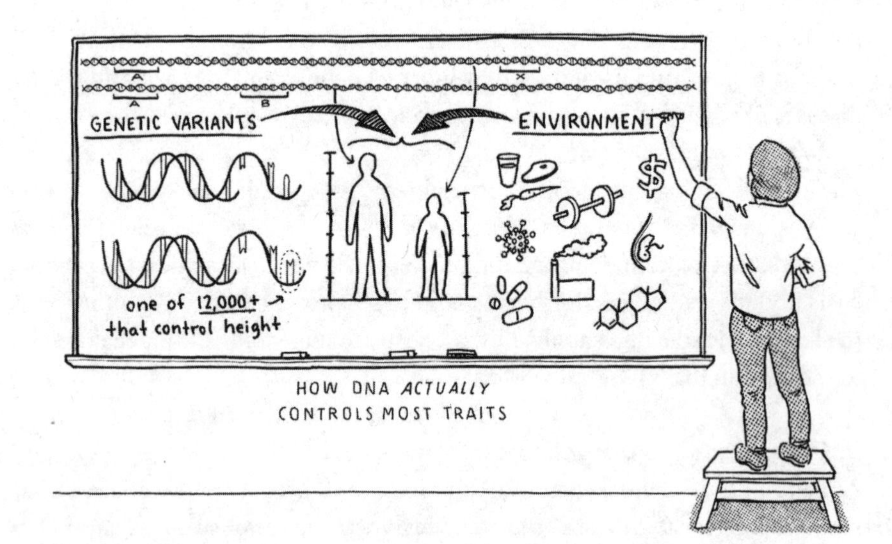

HOW DNA *ACTUALLY* CONTROLS MOST TRAITS

Even more, it turns out that some variants of genes that are associated with traits like height might have effects that rely on the presence of other gene variants. Maybe gene variant A and gene variant B are more present in tall people, but that same gene variant A and gene variant X co-occur in people of average height. We call this interaction between genes epistasis, and it really complicates how we search for simple genetic causes for traits. The story of how genes work isn't about what they do on their own, but often how they work in the presence of other genes.

So yeah, things are pretty complicated. And I regret to inform you that the story might be even messier than I've presented. Because we now know that how genes function is influenced by the setting that they are in. For example, we now understand that being born and raised in resource-poor settings can contribute to making people shorter than those raised in resource-rich settings. Why? Because whatever genes contribute to height, how they do their job depends on nutrition and other features of an organism's environment. This phenomenon—called gene-by-environment interactions—is a powerful idea. It suggests that the only way to understand the full story of how genes make life is to know more about the context in which that organism was raised. That even applies to Mendel and his pea plants—yes, he got a lot of clean answers. But even his results depended on the particulars of the setting the plants were grown in.

3. Why Do We Care?

Are we saying that Gregor Mendel was wrong? Should we change how we remember him? Of course not. Science is a dynamic process, and even the best of us get things wrong. But, honestly, Mendel was not really wrong by most definitions. The better summary is that the truth he uncovered worked for a certain set of traits, in a certain sort of organisms. If there was a mistake, it was ours—we erroneously believed that all of our genes worked like Mendel's pea plant, an assumption we are still dealing with in our modern searches for genes associated with disease. But the big problem with this misapplication of Mendel is not only in the science itself, but in how it was applied.

Once people started to believe that everything worked like a pea plant, we started to erroneously conclude that we could understand everything about how life—even human life—was constructed. So if we could understand the genetic basis behind the color of pea plant leaves, we'd be able to easily understand why some people were rich, others were poor. And this led to the unfortunate era where we made unscientific and harmful claims about who humans were, and especially about our differences. This led to one of the ugliest misapplications of bad science: the eugenics movement, where policies were enacted to actively prevent or discourage people from certain groups to reproduce. I would love to report that this sort of misunderstanding of genetics is a thing of the past. But it remains an obstacle in modern society.

So what can we learn going forward? Well, that the story of DNA has more twists and turns than the double helix structure doesn't take away from the magic of genetics. There remain so many important things to discover about life on Earth that genes play a central role in. And that there might not be a single "stuff of life" makes it more exciting: There are more forces to consider and actors involved, which makes DNA all the more magical.

Dr. C. Brandon Ogbunu is an assistant professor at Yale University and an external professor at the Santa Fe Institute. His research takes place at the intersection of evolutionary biology, epidemiology, and complex systems.

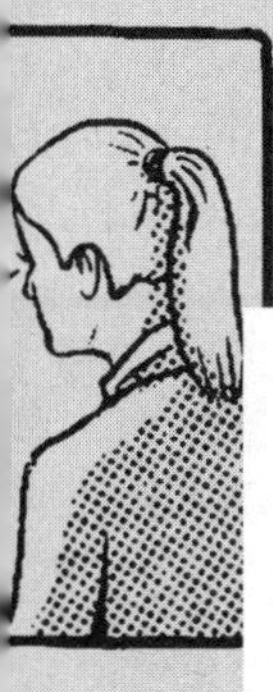

Pathogens and Parasites

This chapter is perhaps being written . . . too soon. It may seem a distant memory at this point, but even before the COVID-19 pandemic, pathogens and parasites wreaked havoc on our species and every other species on Earth. As the parent of a four-year-old, I'm learning firsthand about this onslaught of pathogens. Some of the pathogens we've faced are viral—COVID, the flu, hand, foot, and mouth disease, roseola, and RSV (respiratory syncytial virus); others are bacterial, like strep, pink eye, and common ear infections. One or more members of our family have had all of these in the last few years, and I've produced more different colors of snot than I knew existed. But nobody really wants to hear about that right now. So instead, in this chapter we've got parasitic birds and parasites of birds. Unlike the chapter misleadingly titled "Mmm . . . Brains," this chapter actually *does* include extensive discussion of zombies and zombification. And instead of bacteria and viruses, we've got worms. Here in the US we are fortunate not to suffer extensively from worms, but we wanted to bring them back into the cultural consciousness. Finally, we include a piece on the immune system, that wonderfully adaptable system that, when functioning optimally, can bail us out of a lot of sticky situations.

—Chris

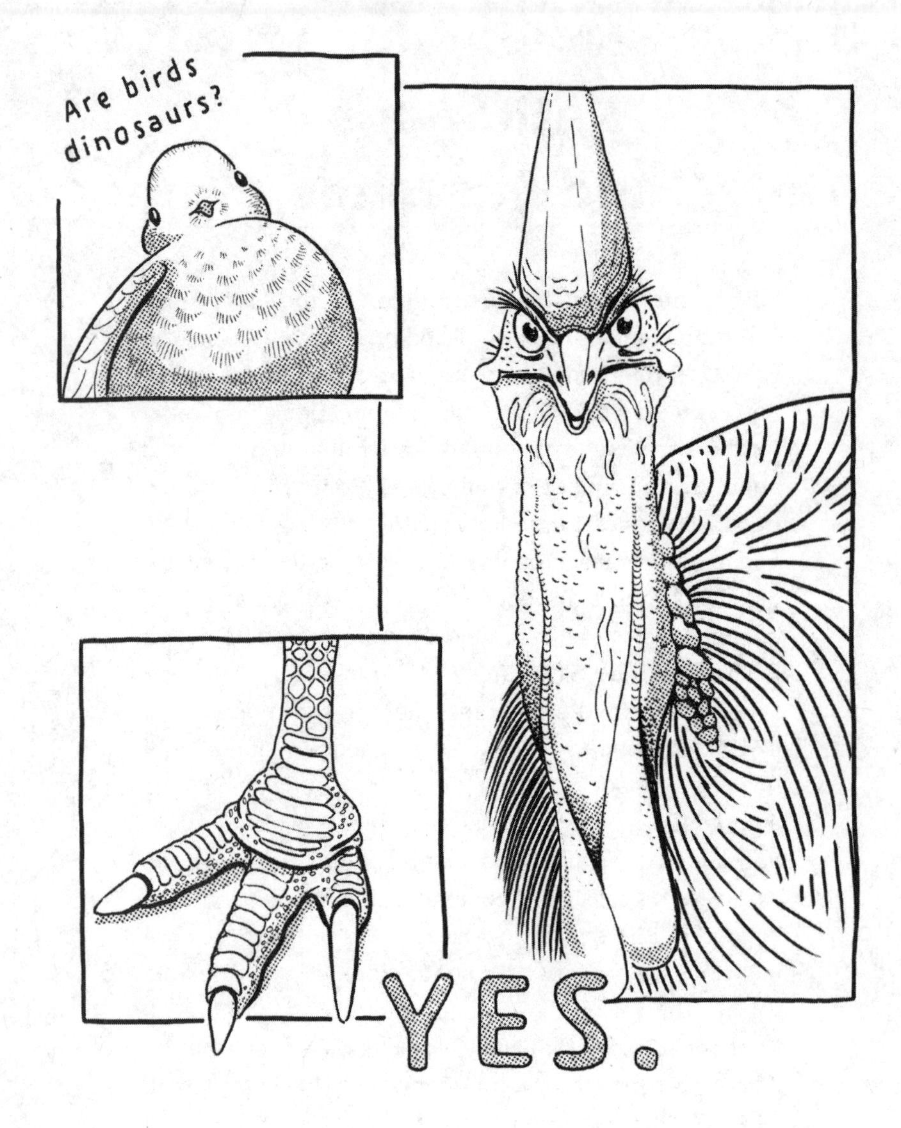
Are birds dinosaurs?
YES.

NERD NITE BOSTON

EVERYTHING YOU ALWAYS WANTED TO KNOW ABOUT BIRDS

by **Dr. Christopher N. Balakrishnan**

I study birds for a living. Yes, I am an ornithologist. Yes, I get paid to look at birds. Why do I study birds? Most notably, I was drawn to birds by their spectacular beauty and their remarkable vocal and courtship behaviors. I was inspired by scientists like Peter and Rosemary Grant, who toiled away on remote islands in order to understand how one species of bird becomes two species of bird. But when I am at a party, do people ask me about these aspects of bird biology? No they do not. What I usually get are the following:

1. Are Birds Dinosaurs?

Yes. All you need to do to answer this question is look at this picture of a cassowary.

Birds are the modern descendants of dinosaurs. This is a fact, not in any way controversial or uncertain. We know that birds are dinosaurs because careful analyses of fossils show more similarities between birds and dinosaurs than with other reptiles. And if dinosaurs and birds are closely related, you can guess what a dinosaur might taste like. Hint: chicken.

2. How Do Birds Have Sex?

I'll admit, even as I started my PhD, this was a little bit confusing to me. I think what confuses folks is that most birds don't actually have an "intromittent organ." Bird sex primarily involves something that we in the business call a cloacal kiss. That is, the male orifice (cloaca) contacts the female orifice (also cloaca), and in a brief moment of beauty, sperm is transferred. There is, however, a major exception to this avian rule of thumb: ducks.

Ducks *do* have an intromittent organ, and it comes in all shapes and sizes. Whereas most birds exemplify the beauty and splendor of romance—males performing elaborate courtship repertoires just to garner a fleeting glance from a female bird—ducks highlight the darker side of sexual reproduction—the battle of the sexes for control of reproduction. Duck penises in some cases have evolved into horrific corkscrew-shaped devices, and female duck reproductive tracts have evolved to corkscrew in the opposite direction to prevent insemination by unwanted male ducks.

3. What Are Parasitic Birds and Why Are They So Amazing?

Well, nobody really asks me this question, but this is something I actually research, so I want to write a bit about it. But no, parasitic birds do not suck blood. Rather, parasitic birds are what we call brood parasites, aka nest parasites. In a very un-birdy fashion, parasitic birds do not build nests at all. Rather, they lay their eggs into the nest of another bird species, the host, and these unsuspecting victims are coerced into raising the parasitic babies.

Brood parasitic behavior has evolved in *seven* different groups of birds. As an analogy, think about flight in vertebrate animals. Birds and bats fly, representing two independent origins of flight. Just within birds, the odd strategy of brood parasitism has evolved seven different times in seven different bird groups (yes, including ducks). This bizarre strategy includes not only skipping the whole nest-building thing but also abandoning any aspects of parental care of offspring, and it requires that parasitic species are able to find suitable host nests in which to lay their eggs.

By far the most famous of the parasites are the cuckoos in Europe (you know, the ones in the clocks), but each of these seven groups of parasitic birds could be thought of as a different "flavor" of brood parasitic behavior. For example, in some cases, like the indigobirds of Africa, parasitism is relatively benign. The baby indigobirds are reared by host parents alongside the host chicks. In a sense, the baby parasitic indigobirds just represent an extra mouth or two to feed. Certainly, providing this extra food takes effort from the host parents, but it is nothing like the bloodbath that is honeyguide parasite babies. Baby honeyguides, as benign as they sound, are born with a hook on their bill and they use this to murder their baby nestmates. There are YouTube videos of this.

Another way in which parasites vary is in terms of how selective they are in where they lay their eggs. Brown-headed cowbirds, native to North America, will lay their eggs basically anywhere—they are known to parasitize over 200 species and have been implicated in the declines of some of these species. Other parasitic species like indigobirds lay their eggs in only a single type of host species nest. In these host-specific parasites, mimicry often evolves, where the babies look exactly like the host chicks, preventing discrimination by the host parent. See, (brood) parasites are amazing and birds are totally real.

Chris founded Nerd Nite in the early 2000s while studying brood parasitic birds in Boston and is quite pleased that it is still a thing (and now available in book form).

NERD NITE WAGGA WAGGA, AUSTRALIA

PIGEONS, CANNIBALS, AND VAGINAS: The Story of My Favorite Parasite

by **Andrew Peters**

POV: I'm a parasite in your vagina but a pigeon's mouth feels like home, too.

Everyone fusses about Aussie parrots, but when it comes to birds Down Under, we have a bit of a hidden secret: Australia and New Guinea have a stunning diversity of pigeons and doves, from desert-dwelling spinifex pigeons and rainbow-colored rain-forest fruit doves to dainty diamond doves and chonky (more than four kilograms!) crowned pigeons. Most of the more than 130 Australasian columbid (fancy name for pigeon and dove) species have been here since we tore away from the rest of Gondwana some thirty-plus million years ago. They are such a distinctive feature of our region that, together with parrots and kingfishers, they were key to the famous naturalist Alfred Wallace recognizing Australasia as a distinct bioregion in 1859.

These birds hit peak diversity in the Australasian tropics, which proved to be rather convenient when a bird-obsessed PhD student like myself was interested in (1) the unique features of Australian–New Guinean fauna, (2) animals that don't bite or scratch, and (3) tropical island and rain-forest-clad paradises. But having first trained and worked as a veterinarian, I was not so much interested in the animals as in what was happening *inside* them.

And this is why one particular microscopic parasite attracted my attention: *Trichomonas gallinae*.

This incredibly common parasite lives in the throat and crop (a deep, expanded section of the throat) of domestic pigeons. It also causes one of the oldest known diseases of birds, the charmingly named canker, which can be lethal in birds of prey when they feed on pigeons. And if *you* happen to have canker, it's almost certainly not caused by *Trichomonas*, unless it's on your genitals and/or you've had disturbingly close contact with the oral cavity of a pigeon (if so, please seek medical advice).

While *Trichomonas gallinae* is well established in introduced pigeons in Australia, at the time of my interest, nothing was known about whether it was found in our fabulous native columbids or whether it posed a risk to them. And as a tragic nerdy conservationist, those kinds of known unknowns give me clammy hands and mild tachycardia. So I set off on a veritable adventure across remote parts of northern Australia and Papua New Guinea to find out what was living in the crops of our pigeons and doves, armed with a trailer load of field equipment and some rum for mojitos.

Dozens of species and many hundreds and of pigeons and doves (caught, sampled, and lovingly returned to their day jobs) later, I had some answers. Of course, these weren't what I had expected.

It turns out that columbids in Australia and New Guinea do very commonly have *Trichomonas*, but not *Trichomonas gallinae*. In fact, they have their own incredibly diverse set of unique *Trichomonas* species! These have probably been living in our pigeons and doves for millions of years and don't appear to cause them any disease. Perhaps, because our columbids already have their own *Trichomonas*, the disease-causing *Trichomonas gallinae* won't be much of a risk to them. Perhaps *Trichomonas* benefits our pigeons and doves in some way, maybe even by killing off birds of prey that target them.

I don't know, but what I do know is that all of the new and genetically diverse *Trichomonas* species in Australasian columbids told us more about the evolution of *Trichomonas* parasites. They appear to have originated in pigeons and doves and occasionally spilled over into other species, which then became new hosts.

And that's where vaginas come into this story.

The most famous *Trichomonas* species of all is *Trichomonas vaginalis*, which makes its home in *our* lower reproductive tract. *T. vag* is a surprisingly common parasite—about one in 50 women in the US are infected, which is a significant health problem causing vaginal discomfort and increasing the risk of HIV transmission in men and women. Perhaps even more surprising than its prevalence is its likely original source—the "deep throat" of pigeons and doves! No, I'm not suggesting that there is modern transmission of *T. vag* between pigeons and humans, but rather a more ancient jump between birds and humans, perhaps similar to what we observed with COVID-19 several years ago. That is where my hypothesizing grinds to a sanitized halt. For more details, bring tequila.

Now the funny thing about adventures is that they can take us to unexpected places. I don't just mean physical places, though that is true enough, too. I mean personal, intellectual places. My adventure to Papua New Guinea did just that, landing me on a different intellectual trajectory than the one I set off on. PNG is an extraordinary place by almost any measure. Hosting globally significant biodiversity both on its rugged landscape and in its breathtaking seas, PNG is equally home to astounding cultural diversity. With over 840 languages, PNG is also one of the most linguistically diverse places on Earth. If two Papua New Guineans meet at random there is only a 1 percent probability that they will speak the same mother tongue! Though undergoing rapid change, PNG is still a place where indigenous knowledge systems and ways of living are the norm.

Every journey to PNG has challenged my understanding of what it is to be human and the relationship between society and the rest of nature. *Trichomonas* isn't the focus of my work anymore. Neither are parasites or pigeons. My PhD adventures instead set me on course to hopefully better understand the relationships among people, wildlife, and our shared health on a deeply interconnected planet.

your/our is used in the general, rather than personal, sense, I hope . . .

Once a vet, always a wildlife lover and outdoors enthusiast, Andrew has ended up somehow becoming associate professor in wildlife health and pathology at Charles Sturt University in Australia and president of the Wildlife Disease Association.

NERD NITE MEMPHIS

WHAT BIRDS CAN TEACH US ABOUT THE IMPENDING ZOMBIE APOCALYPSE

by James S. Adelman

If Hollywood has taught us anything, it's that someday soon, a fast-acting, highly contagious pathogen will start turning humans into zombies, collapsing our civilization into an anarchy of the undead. These are just facts, people!

Clearly, we need to be prepared, and I don't just mean brushing up on your close-combat training and stocking the bunker with an adequate supply of Twinkies. I mean understanding the root cause of this impending apocalypse: your foe and mine, the inevitable zombie virus. How is it most likely to emerge? And once it's among us, how's it going to evolve? Yes, evolution's a thing, deal with it.

Luckily for humanity, bird-nerd scientists (and others, to be fair) have been asking such questions for years now. Not about the zombie virus (yet!) but about some parallel issues worth exploring.

First, let's talk about how new pathogens find their way into humans. As a nature lover and wildlife advocate, I hate saying this, but new diseases disproportionately come from animals (maybe there's a reason it's called bird flu?). For example, a global analysis of emerging infectious diseases—totally new pathogens or old ones that are increasing in frequency—found that about half originated in either domestic or wild critters.[1] Now, to be clear, a big reason such things happen is that we keep building human habitats over natural ones

1 K. E. Jones et al., "Global Trends in Emerging Infectious Diseases," *Nature* 451 (2008): 990–94.

and housing our domestic animals at ridiculously high densities, sometimes very close to wildlife.[2] Also, people need to stop making out with their chickens. So, blame the people, not the critters.

Okay, so a lot of infectious diseases come from animals, but which critters should we worry about most? Well, birds can teach us some interesting, sometimes counterintuitive lessons. For brevity, let's stick with one well-known example, West Nile virus. Normally, West Nile is like a cold or flu in humans, but for vulnerable folks, it's no joke, having caused over 50,000 cases and 2,400 deaths since its introduction to the US in 1999.[3] Now, you may not remember the late 1990s and early 2000s (too young, too interested in party drugs at the time, whatever, I'm not judging), but tons and tons of crows were dying from West Nile across the US.[4] So you may be thinking, "I've seen Hitchcock, I know to watch out for crows in general, and if they're all dying of something, that goes double, right?"

Let's think about that. Sometimes dead animals can spread nasty pathogens (anthrax, for example),[5] but for West Nile virus, a dead bird is a dead end. The virus needs a mosquito to bite a living bird so it can hitch a ride to the next animal that that mosquito bites, be it a bird or a human. So what we really need to worry about are birds that carry the virus, survive quite well, and look absolutely mouthwatering (proboscis-watering?) to mosquitoes. With this in mind, researchers found that crows weren't the main cause for concern, it was the American robins: They're pretty common, do okay when infected, carry a good amount of virus, and are freaking irresistible to the mosquitoes that can carry West Nile.[6]

So, lesson 1 for our zombie future: Don't get too distracted by big animal die-offs. I mean, get concerned—like really concerned—but also know there could be a silent killer out there carrying zombie virus and doing just fine. (We're

---

2 T. Allen et al., "Global Hotspots and Correlates of Emerging Zoonotic Diseases," *Nature Communications* 8 (2017): 1124.

3 Centers for Disease Control and Prevention, "West Nile Virus," http://www.cdc.gov/westnile/index.html (accessed on November 18, 2022).

4 A. M. Kilpatrick, S. L. LaDeau, and P. P. Marra, "Ecology of West Nile Virus Transmission and Its Impact on Birds in the Western Hemisphere," *Auk* 124 (2007): 1121–36.

5 M. Hugh-Jones and J. Blackburn, "The Ecology of *Bacillus anthracis*," *Molecular Aspects of Medicine* 30 (2009): 356–67.

6 Kilpatrick, LaDeau, and Marra, "Ecology of West Nile Virus Transmission"; A. M. Kilpatrick et al., "Host Heterogeneity Dominates West Nile Virus Transmission," *Proceedings of the Royal Society Biological Sciences* 273 (2006): 2327–33; A. M. Kilpatrick, "Globalization, Land Use, and the Invasion of West Nile Virus," *Science* 334 (2011): 323–27.

pretty sure that's what happened with COVID and bats,[7] so to all the politicians who will in no way whatsoever read this, please don't shut down our wildlife disease surveillance programs.)

Now, when the zombie virus eventually spills over from its animal reservoir, how can we expect it to evolve? Again, birds have taught us about something pretty scary: Sometimes newly emerged pathogens just keep getting nastier—they evolve increasing virulence. Virulence is basically all the bad shit that happens while you're infected, up to—or, in the case of the zombie virus, beyond—death.

Back to the birds. I'm thinking of lessons learned about virulence from house finches and conjunctivitis (like pink eye in people but from a totally different pathogen). In the mid-1990s, folks detected a bacterial pathogen that jumped out of poultry into wild birds, most notably house finches. It quickly spread across North America, causing such bad lethargy (sleepy birdies) and conjunctivitis that infected birds got worse at avoiding predators[8] and finch populations plummeted by up to half.[9] Curiously, although theory might suggest that the pathogen should've evolved to an intermediate level of virulence (don't get too, too nasty or else you'll kill your host before it can pass the infection to others),[10] over the following decades that's not what happened. Once it was established in a location, the bacteria kept getting nastier, causing worse and worse eye swelling,[11] which, in this disease, helps it escape from bird immune systems[12] and helps infected hosts keep infecting new individuals.[13]

After all that, what are the big take-home messages of all this bird stuff for our zombie virus?

---

7 S. Lytras et al., "Exploring the Natural Origins of SARS-CoV-2 in the Light of Recombination," *Genome Biology and Evolution* 14 (2022): evac018.

8 J. S. Adelman, C. Mayer, and D. M. Hawley, "Infection Reduces Anti-Predator Behaviors in House Finches," *Journal of Avian Biology* 48 (2017): 519–28.

9 A. A. Dhondt, A. P. Dobson, and W. M. Hochachka, "Mycoplasmal Conjunctivitis in House Finches: The Study of an Emerging Disease," in *Wildlife Disease Ecology: Linking Theory to Data and Application*, ed. K. Wilson, A. Fenton, and D. Tompkins (Cambridge: Cambridge University Press, 2019).

10 S. Alizon et al., "Virulence Evolution and the Trade-Off Hypothesis: History, Current State of Affairs and the Future," *Journal of Evolutionary Biology* 22 (2009): 245–59.

11 D. M. Hawley et al., "Parallel Patterns of Increased Virulence in a Recently Emerged Wildlife Pathogen," *PLoS Biology* 11 (2013): e1001570.

12 A. E. Fleming-Davies et al., "Incomplete Host Immunity Favors the Evolution of Virulence in an Emergent Pathogen," *Science* 359 (2018): 1030–33.

13 C. Bonneaud et al., "Experimental Evidence for Stabilizing Selection on Virulence in a Bacterial Pathogen," *Evolution Letters* 4 (2020): 491–501; R. M. Ruden and J. S. Adelman, "Disease Tolerance Alters Host Competence in a Wild Songbird," *Biology Letters* 17 (2021): 20210362.

1. Look for the zombie virus in animals, but not necessarily in the sickest animals.

2. Be concerned that once it jumps into humans, it could get even nastier.

In this case, getting nastier might look like a virus that is more lethal and makes faster and faster zombies. So the original strain might produce slow-to-emerge-from-their-graves, shambling zombies that look like Romero's *Night of the Living Dead*, but while you're eating Twinkies in your bunker, keep an eye out for rapidly reanimating, fast-moving zombies with more of a *World War Z* feel.

Now that we're armed with that knowledge, hopefully we'll stay a little safer out there as the world ends. Also, don't forget your cardio.

Jim Adelman is an assistant professor at the University of Memphis whose research, unsurprisingly, focuses on diseases in wild birds.

NERD NITE BOSTON

ZOMBIES ARE REAL AND YOU MIGHT BE ONE

by Jeremy N. Kay, PhD

Folks, I hate to be the bearer of bad news. But I'm here to tell you: Zombies are real. No, not *Walking Dead*–style zombies. But have no doubt: Actual zombies walk among us. Creatures whose brains have been taken over—or dare I say eaten?—by the dark, terrifying forces of the natural zombie world.

And what forces might these be? Please allow me to introduce *Microphallus turgidus*. Yes, its name means more or less what it sounds like.

Microphallus turgidus and its trematode relatives are little worms that are ready to invade your cranium and take over your brain—if you're a freshwater shrimp. Why would they do this? *Microphallus* worms are parasites that live in the guts of waterfowl. Cozy in their happy intestinal home, they lay a bunch of wormy eggs; these in turn get pooped out into the water, where they settle to the bottom and get eaten by an unsuspecting shrimp. Soon the egg hatches and the larva takes up residence inside the shrimp. But now, baby *Microphallus* has a problem: It turns out that shrimp prefer to hang out at the muddy bottoms of lakes and rivers, where there is plenty of shrimp food and they are safe from predators such as waterfowl. How, then, is poor baby *Microphallus* supposed to return to its happy intestinal home where it can become a grown worm and reproduce? I'll tell you how: By taking over the shrimp's brain and making it get eaten by a bird.

After *Microphallus* gets done messing with the serotonin system of the shrimp brain, among other things, the infected shrimp no longer enjoys the nice, dark, safe confines of the muddy lake bottom. Instead, it prefers to swim toward bright light, which leads it to hang out at the surface of the lake and become easy pickings for the neighborhood duck. And voilà! *Microphallus* has completed its life cycle, thanks to the aberrant wanderings of a zombie shrimp.

Okay, I know what you're thinking. You are not a shrimp, so what's the big deal? Well, it turns out that the *Microphallus* genus isn't alone: There are lots of other parasitic worms that can pull similar tricks, taking over their larval hosts' brains to deliver them into the bellies of the creatures where they can complete their life cycle. And it doesn't stop with worms: There are vast numbers of parasitic wasps that use their stinging venom to take over the brains of their insect hosts.

Sometimes the venom just paralyzes the zombie insect, turning it into a living, breathing food source for the wasps' babies. But some zombie behaviors are more elaborate, such as the cockroaches who are willingly led by the antenna back to a wasp's den to become larval wasp dinner or the zombie caterpillars that use their bodies as battering rams to fend off any predators that try to eat the wasp's cocoons. Even fungi can get in on the fun: Zombie insects infected by one of these fungal species will climb up to the highest possible point on the nearest plant stalk, where they remain until their death. At that point, fungal spores begin to rain down from the elevated insect corpse, thereby dispersing the spores over a larger area than if the insect had remained at ground level.

In case you're still not feeling too threatened by the zombies that walk among us, we still need to meet one more character: *Toxoplasma gondii*. These are single-celled creatures that reproduce in the guts of cats. After their immature "spore" form gets deposited in your litter box, it may make its way into the body—and ultimately the brain—of a small rodent. And dear reader, can you guess how this turn of events affects the way the rodent may feel about its feline predators? Let's just say, the *Toxoplasma* life cycle is a lot easier to complete when a mouse's fear of being eaten is diminished.

Now as it turns out, *Toxoplasma* can make its way into the brains of most mammals, *including humans*. In fact, as many as 25 percent of people harbor a *Toxoplasma* infection. Our immune systems are generally quite good at dealing with these creatures, so they don't cause widespread damage—or even any symptoms at all in most infected mammals. But there's some evidence to suggest that behavioral changes may ensue. Like rodents, infected wolves show evidence of greater risk-taking, as well as behaviors that bring them in closer proximity to cougars (a feline *Toxoplasma* host). In humans the evidence for altered behavior is thinner, but I would simply note that a very large fraction of people choose to share their homes with a certain species of fluffy zombie-making fur baby. So, if you ever thought the "crazy cat person" was a made-up trope, that character is real and walks among us today! Be careful the next time Matt invites you over to his house.

Jeremy Kay is former boss of Nerd Nite Boston. He is currently associate professor of neurobiology at Duke University. These days he mainly nerds out about the retina, but zombies will always be near and dear to his heart/brain axis.

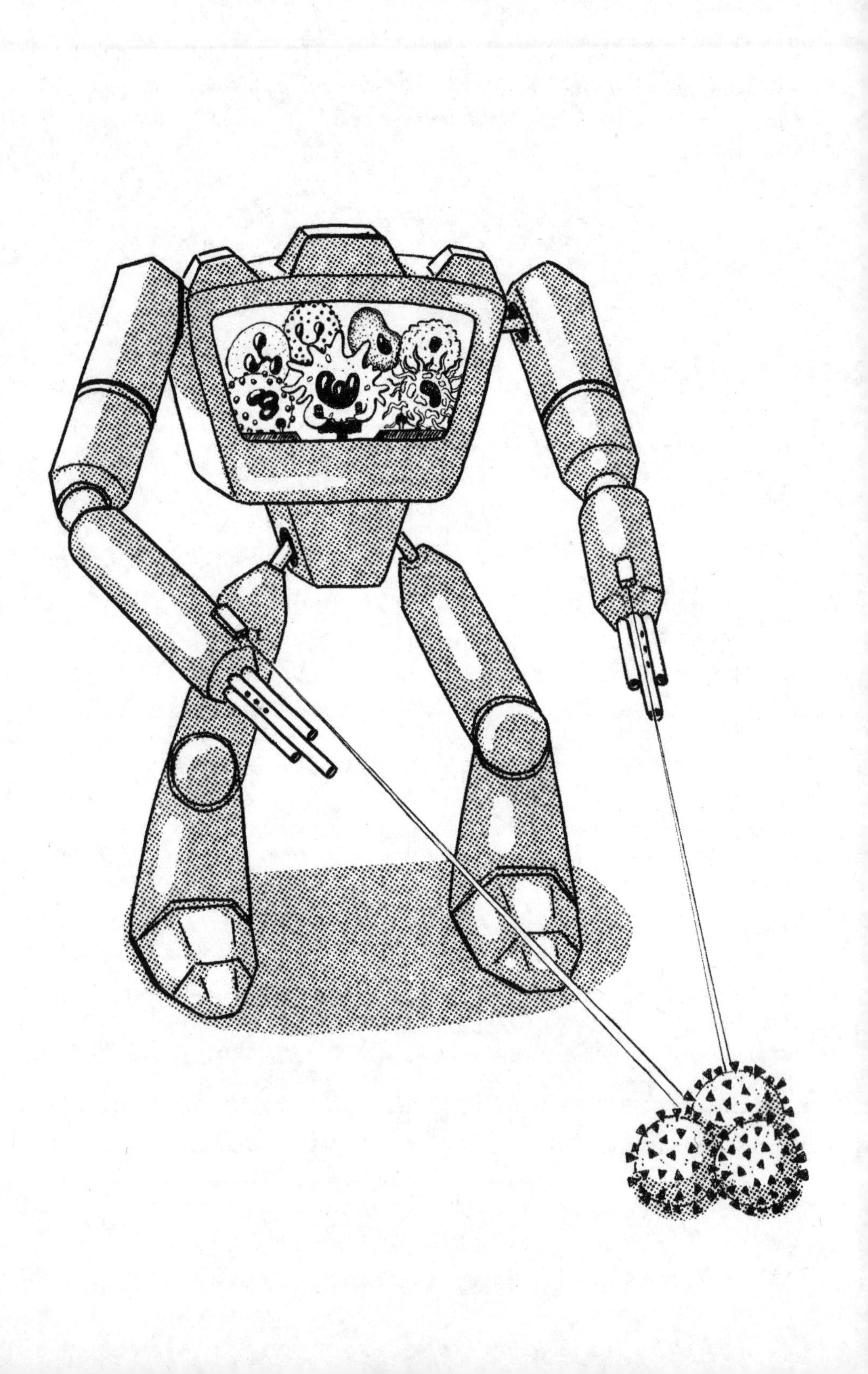

NERD NITE MELBOURNE

HACKING THE ANTIVIRAL IMMUNE RESPONSE

by **Dr. Ebony Monson**

Several years after the height of the COVID-19 pandemic, many people know the telltale symptoms of a COVID-19 infection: a sore throat, a pernicious cough, congestion, fever, and full-body exhaustion. In fact, from the start of the pandemic through November 2022, SARS-COV-2, the virus responsible for the COVID-19 pandemic, had infected more than 600 million people worldwide and sadly killed 6.5 million, according to the World Health Organization. Unfortunately, this is not the worst pandemic we have seen, and it's predicted it won't be the last. Yet with all this attention and focus on COVID-19, not many people understand what a virus actually is and how your body reacts once you are infected.

So what is a virus?

A virus is an infectious microbe consisting of a segment of nucleic acids (either DNA or RNA) surrounded by a protein coat. In the case of respiratory viruses such as SARS-COV-2 or influenza, these non-living entities infect us by entering through our mouth or nose and then ultimately go on to hack our cells, turning them into machines for producing more copies of themselves. But what do our cells do to combat this infection? After a virus infects our cells, it doesn't take long for our immune system to detect and start fighting off these invaders. Our extremely complex immune response is very robust, and in general works well to clear viral infection, but viruses have evolved strategies to combat our immune response, and that's when we usually get sick.

Our body's response to infection can be broken down into two stages. First is the innate immune response, which is activated immediately following infection, producing antiviral molecules that combat infection. And second, the adaptive immune response is activated later during infection and is responsible for "remembering" past infections. The second stage of the immune response is the most well known as this is the stage of your immune response that vaccines support. We all know that vaccines have been widely successful in the eradication of viruses such polio and smallpox, but during the COVID-19 pandemic we also learned how expensive and time-consuming they are to develop. Additionally, they are also specific to one virus and are often strain-specific, meaning that if the virus mutates, it will make that specific vaccine less effective.

As the COVID-19 pandemic persists, scientists around the world are working tirelessly to outsmart new variants and plan for the next virus with pandemic potential. One way in which we can do this is by developing antiviral drugs targeted to our own immune response, instead of targeting the virus itself. That is, manipulating the first innate stage of our immune response, which is activated immediately following infection.

This means we can boost our immune response so that it is able to fight off the virus infection! Amazing, right?

Drugs that can do this will be effective against multiple different viruses, because they initiate a broad antiviral response but don't target specific strains of the virus itself. This means other infections that currently have no good antiviral therapies will be treatable! So stay tuned for these new antivirals, because the future in this space is looking bright!

Dr. Ebony Monson is a postdoctoral researcher at La Trobe University in Melbourne, Australia. She completed her PhD in 2021, and her research is focused on the development of novel antiviral strategies against multiple viral infections. She is passionate about science communication and mentoring the next generation of women in STEM!

NERD NITE NYC

HUMAN PARASITES (NO, NOT YOUR MOOCHING ROOMMATE)

by **Dr. John Dodson**

We're going to talk about worms because they're fun.

In medical school, even if a student knows exactly the field in which they wish to specialize, they still have to learn plenty of wide-ranging biology, anatomy, and physiology. From cell structure to trauma treatment, the breadth and depth of information is vast. And for me, parasitology was one of my favorite, yet least relevant, courses. Least relevant because, by and large, most parasitic infections have been eliminated from the US. And favorite because the images and ghastly stories will always draw oohs and aahs from my cringing friends and family. Nonetheless, human parasites are fascinating, and they still affect billions of people worldwide.

There are numerous parasites that infect humans, ranging from large roundworms (*Ascaris lumbricoides*) to the dreaded brain-eating amoeba (*Naegleria fowleri*), but ascaris is particularly clever. And by clever, I mean dastardly. These roundworms are essentially large blind worms that hatch in the human intestine and live on whatever a person (the host) eats. And they affect an estimated *one billion* people worldwide. That's one in seven humans.

Fortunately, they're easily treatable with medications, but for people who are untreated and/or don't have access to the proper medication, the worms can multiply and ultimately block anything from getting through, thereby creating an intestinal obstruction. Yes, a worm can prevent anything from passing through you.

Which brings me to the aforementioned brain-eating *Naegleria fowleri*, which usually infects people when the water in which it lives enters a human through their nose. This typically happens when people go swimming, dive, or put their heads under fresh water. Yep, just like in that small inland lake or pond you go skinny-dipping in every summer. As the CDC website nicely describes, the amoeba then travels up the nose to the brain, where it destroys the brain tissue and causes a devastating infection called primary amebic meningoencephalitis (PAM), which is "almost always fatal." Fortunately, chlorine in pools and water parks eliminates this parasite, and even if chlorine levels are too low, it'll still mitigate any fatal risk. And for all of you who use neti pots to clean out your noses every winter—watch out—contaminated tap water can also harbor the amoeba. Nothing like trading some mucus for a brain-eating beast!

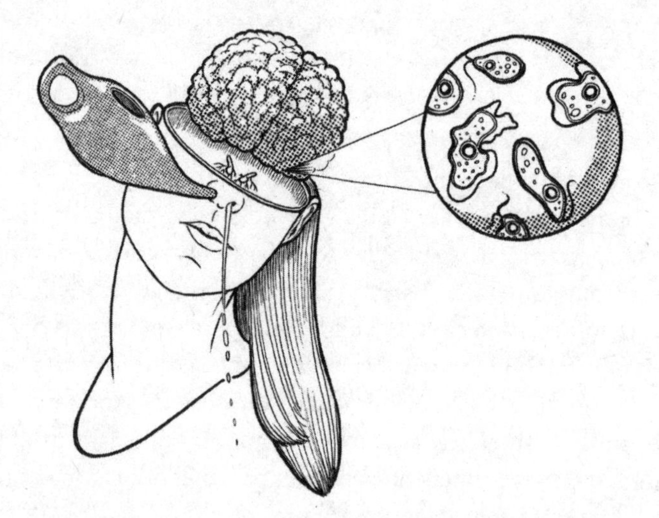

Finally, then there's the pinworm, a microscopic worm that also lives in human intestines but can crawl out of the rectum at night to lay eggs. This phenomenon is famously diagnosed via the "Scotch tape test," which is conducted by putting tape near a human's rectum to pick up eggs and look at them under a microscope. These worms are most common in children and are quite treatable if they're detected—and they're usually detected when a person develops itching where the worms crawl out.

John Dodson is a cardiologist who works in New York City.

Death and Taxes (But Really, Just Death)

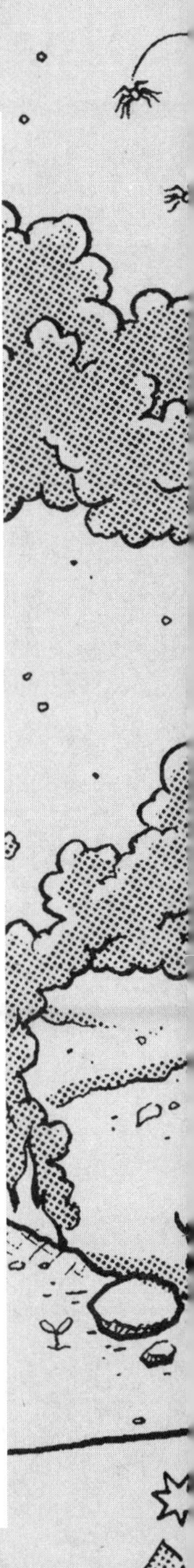

At Nerd Nite, we pride ourselves on making science fun. So here we present the death chapter. Fun! Fun? I can't say that when we were brainstorming about this book I envisioned a chapter about death, but it turns out, of course, that death is the great unifier. This chapter spans topics touched on in other parts of the book: astronomy, fecal matter, genetics, psychology, biology, but all wrapped up into a cheery package surrounding death. Curious about how tiny organisms called cyanobacteria killed a whole lot of life on Earth or about other paths to mass extinction? Then this is the chapter for you. In this chapter you will also learn about some things that pertain more directly to your daily life (I mean death)—like the basics of how cancer works. Fear not, it is not all doom and gloom. This chapter also tells the heartwarming tale of the life (and death, and taxidermy) of the bear on the California state flag, and draws from science fiction and psychology to tell us how to avoid another mass extinction event. Is tarot card reading still a thing? I have a distinct memory of a tarot card reader in New Orleans explaining to me that drawing the Death Card was not a bad thing since the card symbolizes a new beginning—just like a mass extinction. Hooray!

—Chris

NERD NITE SAN FRANCISCO

MONARCH THE BEAR: A Tale of Tycoons, Taxidermy, and the California Flag

by Kelly Jensen

In November 1889, a train bearing a live grizzly bear rolled into the depot of San Francisco. That bear would become the last California grizzly in captivity and would later be immortalized as the bear on the California flag. This is the story of Monarch the bear.

California grizzlies were a subspecies of the grizzlies that are still around today, but generally characterized as larger and tougher than other grizzlies. There were tons of them throughout California, from north to south, in every habitat from seashore to mountains. If you see place-names with *bear* in them through the state (Bear River, Big Bear, Los Osos), those are generally named for grizzlies rather than black bears.

However, we don't know much about California grizzlies because they were hunted to extinction within 60 years of the Gold Rush in 1849. The last one was shot in Fresno County in 1922, so most of what we know scientifically comes from museum specimens. California grizzlies were most often killed for interfering with livestock and humans, for entertainment (bear-baiting), and pretty much just for funsies. That makes California the only state with a state animal that we extincted ourselves. *cue ironic cheers*

The last California grizzly in captivity was Monarch, who lived in San Francisco from 1889 until 1911.

Monarch was captured at the behest of William Randolph Hearst, the well-known newspaper publisher and [fill in your own early-twentieth-century millionaire playboy noun here]. Best known as the model for *Citizen Kane*, Hearst did things like starting yellow journalism and getting the US embroiled in the Spanish-American War, so . . . yeah. Hearst had a bet about whether California grizzlies were extinct or not, so he sent one of his reporters, Allen Kelly, on a mission to capture a live grizzly and bring it back to San Francisco as a publicity stunt for his newspaper.

I'm going to spare you the details of Monarch's capture, but if you really want to know, you can read Kelly's book, *Bears I Have Met—And Others*. Basically it involved a lot of chains, a sled pulled by some very freaked-out horses, and eventually a train to San Francisco.

When the bear was brought back to the city, Hearst gave him the name of Monarch since his newspaper, the *Examiner*, was known as "the Monarch of the Dailies." He presented Monarch as a gift to the city of San Francisco, whose officials were basically like, "Wowww, no, there is no damn way we're taking a live grizzly bear."

So Monarch ended up at Woodward's Gardens, an amusement park that used to take up two square blocks of San Francisco's Mission District. Opened in 1866, Woodward's Gardens boasted the first aquarium on the West Coast, art galleries, live entertainment, some astonishingly bad taxidermy, and hot-air-balloon rides that were only *occasionally* fatal. Monarch was the star attraction and lived there until the amusement park closed in 1891. Monarch was then sold for $32 to Golden Gate Park, where a menagerie was being established, and would live in the park's bear pits near the kangaroos and bison. Unfortunately, the menagerie was rather meager, but in a positive twist of fate, it inspired the creation of the San Francisco Zoo because its founder saw the animals in Golden Gate Park and wanted to take better care of them.

So what would Monarch have been like if you had gone to see him at the park?

First, he was nearly black, which was very unusual for a grizzly. And he might have been trying to escape so he could eat the moose in the next cage over, in which case there'd have been keepers with iron bars pushing him back into his cage. Or you might have seen him lying around sadly, literally, because he, unsurprisingly, showed signs of depression. When that happened, Hearst stepped in again and arranged to bring Monarch a grizzly lady friend from Idaho. The pair had two cubs in 1905 who lived in a separate cage with their mom. Monarch spent 22 years in captivity before he was euthanized in 1911 due to old age. Eleven years after that, his subspecies was extinct.

After Monarch died, his skin was stuffed and went to the de Young Museum in Golden Gate Park and his bones went to the Museum of Vertebrate Zoology at the University of California–Berkeley; thus, his remains are in two places. Now, when I say Monarch's skin was stuffed, I mean *stuffed*. Badly. Like a couch. He looked tubby and bulgy and mortally goofy. But by 1953, he had been remounted more realistically and had ended up at the California Academy of Sciences, where he was used as the artist's model for the bear on the California state flag.

The now iconic flag was standardized in 1953, but before that, the flag's appearance was a bit, ah, open to interpretation. The typography was weird and the drawing of the bear itself ranged from creepy to cartoonish, so in 1953 the state was like "Oh my God you guys, just MAKE IT LOOK LIKE THIS." Donald Graeme Kelley, an artist at the academy, used Monarch as his model, but based the pose on an iconic Gold Rush image of a grizzly bear by Charles Nahl.

These days, Monarch himself resides in the big taxidermy freezer at the Academy of Sciences. He used to be on display but the light faded his fur pretty badly so he's not out anymore. If you see him now, his fur is reddish brown instead of black and his muzzle is nearly bare from all the folks who petted him when he was on display at the de Young.

As a scientific specimen, Monarch is tremendously valuable because his legacy is evolving even now: The University of California, Santa Cruz paleogenomics lab is sequencing his DNA to learn more about the California grizzly subspecies.

Monarch didn't get the life he should have had, but even in death he reminds us of the wild neighbors humanity has lost and hopefully inspires us to take more care in the future.

Kelly Jensen is your average nerd-about-town. She is a founding fellow of Odd Salon, a photo archivist, and a bestselling author. She likes books, coffee, and cemeteries.

NERD NITE NYC

HOW TO NOT DESTROY OURSELVES:
Lessons from Sci-Fi

by **Dr. Ali Mattu**

The future seemed bright.

As a kid I loved my Atari 2600. Then the Nintendo came out. IT WAS SO MUCH BETTER! I never picked up the Atari again. Same thing happened with the SNES. Then again with the PlayStation.

And more awesome stuff was on the way! The Information Superhighway promised to disrupt the way we communicate and learn through some type of interactive TV. *Back to the Future: Part 2* convinced me hoverboards will be the next big thing. After watching *Contact*, I believed SETI would communicate with aliens in the new millennium (if we survived the Y2K bug, of course).

Humanity was also solving some big problems. In school I learned about the 1987 Montreal Protocol.[1] This was a treaty signed *by the entire world* to fix the hole in our ozone. And . . . it worked! We got rid of CFCs, the chemicals destroying our atmosphere. The ozone hole is healing, saving millions from skin cancer.

Lots of space stuff, too. After the Soviet Union fell, America and Russia announced a partnership to create an International Space Station. Maybe I'd see a joint mission to Mars in my life . . . It's around this time I first saw *Star Trek VI: The Undiscovered Country*. It's a story about Captain James T. Kirk and the Klingons, mortal enemies, making peace with each other. The end is a bunch of different species applauding galactic cooperation. AND I ATE IT UP! When-

1 https://www.unep.org/ozonaction/who-we-are/about-montreal-protocol.

ever I felt down, I'd imagine myself there on the bridge of the *Enterprise*, next to Kirk, helping bring peace to the galaxy.

These events, real and fiction, led me to believe humanity is improving and technology would help us improve faster. I held on to this belief until 2020.

On September 9, 2020, smoke from the North Complex Fire blanketed my home and the San Francisco Bay Area, making our sky look *Blade Runner* orange and our outside air unbreathable.[2] Though the fire was ignited by a lightning storm, it was fueled by decades of bad political decisions and worsened by climate change. And because of indoor COVID-19 restrictions at that time, there was literally no place I could go. The fires might have been unavoidable, along with the COVID outbreak, but we shouldn't have been this helpless and unprepared.

Then came the January 6, 2021, insurrection attempt at the United States Capitol, which felt more like Terry Gilliam satire than, you know, real life. In 2022, Russia invaded Ukraine, and a few months later Russia announced it was leaving the International Space Station.[3] A lot of the *big* progress of my childhood was quickly unraveling. Meanwhile, all that technology I dreamed about as a kid arrived and . . . it mostly sucked. Hoverboards came in the form of self-balancing scooters but were banned because their batteries sometimes exploded.[4] Mark Zuckerberg created a virtual reality metaverse and it looked worse than 1992's *The Lawnmower Man*.[5] We didn't have "true AI," but internet bots and deepfake propaganda were just as effective at creating havoc as the skinjobs on *Battlestar Galactica*.

I became obsessed with Robin Hanson's "great filter" theory. Maybe once life becomes smart enough to explore space, it also becomes powerful enough to destroy itself. That's why we haven't made first contact—no species has survived that challenge. Were we next?

Recently I was watching *Dr. Strangelove* and was struck by how well it captured current global problems. An isolated man believes the fake news that Russians are using fluoride to turn Americans into communists. He issues an order to nuke Russia. Leadership isn't functioning well enough to stop him. Technology malfunctions. Nuclear war!

---

2 https://www.newsweek.com/bay-area-orange-skies-blade-runner-2049–1530961.

3 https://www.npr.org/2022/07/26/1113683450/space-station-iss-russia-leaving-2024.

4 https://time.com/4123076/hoverboard-new-york-illegal/.

5 https://www.forbes.com/sites/paultassi/2022/08/17/does-mark-zuckerberg-not-understand-how-bad-his-metaverse-looks/?sh=129e774037d4.

The film highlights how much we're wired for tribalism. Psychologists have shown that we like the groups we belong to, even when we're randomly assigned to them, and dislike other people's groups for no reason (in-group / out-group bias). Keep groups separate long enough and their views get more extreme (group polarization). Groups without diversity of perspectives don't function well (groupthink). And when our views are challenged, we find ways to keep believing whatever we want (cognitive dissonance, motivated reasoning).

TL;DR: "I like us, dislike them, we're right, they're wrong!"

Americans *have* become more tribal when it comes to where we live, what we watch, and what we believe.[6] The internet seems to have accelerated this, helping us to find whatever ideas support our beliefs and create communities around that. At the same time, globally, we're seeing internet-fueled nationalism with Brexit in the UK, the election of Jair Bolsonaro in Brazil, and Narendra Modi's rule in India. In cases like the genocide of Rohingya Muslims in Myanmar, we see the deadly connection between hateful content on Facebook and violence IRL.[7]

This can't end well. Is it *our* last hurdle, leading to extinction or interstellar exploration?

When I originally gave this Nerd Nite presentation in 2017, the world was a little less messy. But the things I recommended then to not destroy ourselves (abolishing gerrymandering, more humane technology, new perspectives in the political parties) aren't happening. So let me once again seek inspiration from Captain Kirk.

What made him heroic wasn't his twenty-third-century technology. It was how he worked with others to solve problems. He assembled a crew with a variety of backgrounds and always encouraged them to speak their minds. Watch *Star Trek VI: The Undiscovered Country* and you'll notice Kirk was originally against helping the peace effort and wanted to let the Klingons die. Hearing from Spock and Klingon chancellor Gorkon changed his mind and set him on a course toward peace.

We're all anxious, sad, and angry about the state of the world, but humanity can persevere. Maybe it's not about overcoming our tribalism, but embracing it. Like Kirk, we can use our emotions to generate dialogue about the real prob-

---

6 https://www.washingtonpost.com/news/wonk/wp/2015/01/07/the-top-10-reasons-american-politics-are-worse-than-ever/.

7 https://www.theguardian.com/technology/2021/dec/06/rohingya-sue-facebook-myanmar-genocide-us-uk-legal-action-social-media-violence.

lems in front of us—not global issues like melting ice caps, pandemics, or fake news. But making sure our homes, neighbors, and communities are equipped to deal with all the hurdles coming our way.

Captain Kirk didn't believe in no-win scenarios. Why should we?

Ali Mattu is a clinical psychologist. He used to treat anxiety at Columbia University. Then he helped build a start-up that failed. Now he shares what he knows on YouTube.

NERD NITE ALPINE, TX

MASS EXTINCTION

by **Dr. Anirban Bhattacharjee, Thomas A. Shiller II, PhD, and Dr. Sean Graham**

There is a famous Douglas Adams quote: "In the beginning the Universe was created. This has made a lot of people very angry and been widely regarded as a bad move."

Our Earth came into existence 4.5 billion years ago. Early Earth was not an easy place for life to survive. It was devoid of oxygen and constantly bombarded by asteroids. All hope was not lost; there was a silver lining—in the form of an iron nickel core—at the center of our Earth. This core was vital to the development of life, as it prevented Earth from getting bombarded by harmful solar radiation. Eventually plants appeared, contributing oxygen and an ozone layer to the atmosphere, which stabilized things even further. Things were good for a long time. But the universe had other plans.

Mass extinctions happen when at least half of all species die out in a relatively short time. There have been just five such extinction events in Earth's history. The first occurred at the end of the Ordovician period, more than 400 million years ago, during a massive glaciation event. Shortly after the first land-dwelling vertebrates evolved in the Devonian, a second and more enigmatic mass extinction happened. Some have suggested that the reason was a high-energy gamma ray burst that wiped out Earth's ozone layer exposing life-forms to extremely dangerous ultraviolet radiation.

The largest mass extinction in Earth's history took place at the end of the Permian period. Possible culprits for this extinction range from widespread volcanism to meteorite impacts. Regardless, the end-Permian extinction universally affected plants and animals on land and in the sea. The end of the Triassic period is marked by another mass extinction event, chiefly affecting animals on land. The dinosaurs took advantage of the situation and became the dominant group of land-dwelling animals thereafter, until something big happened 135 million years later.

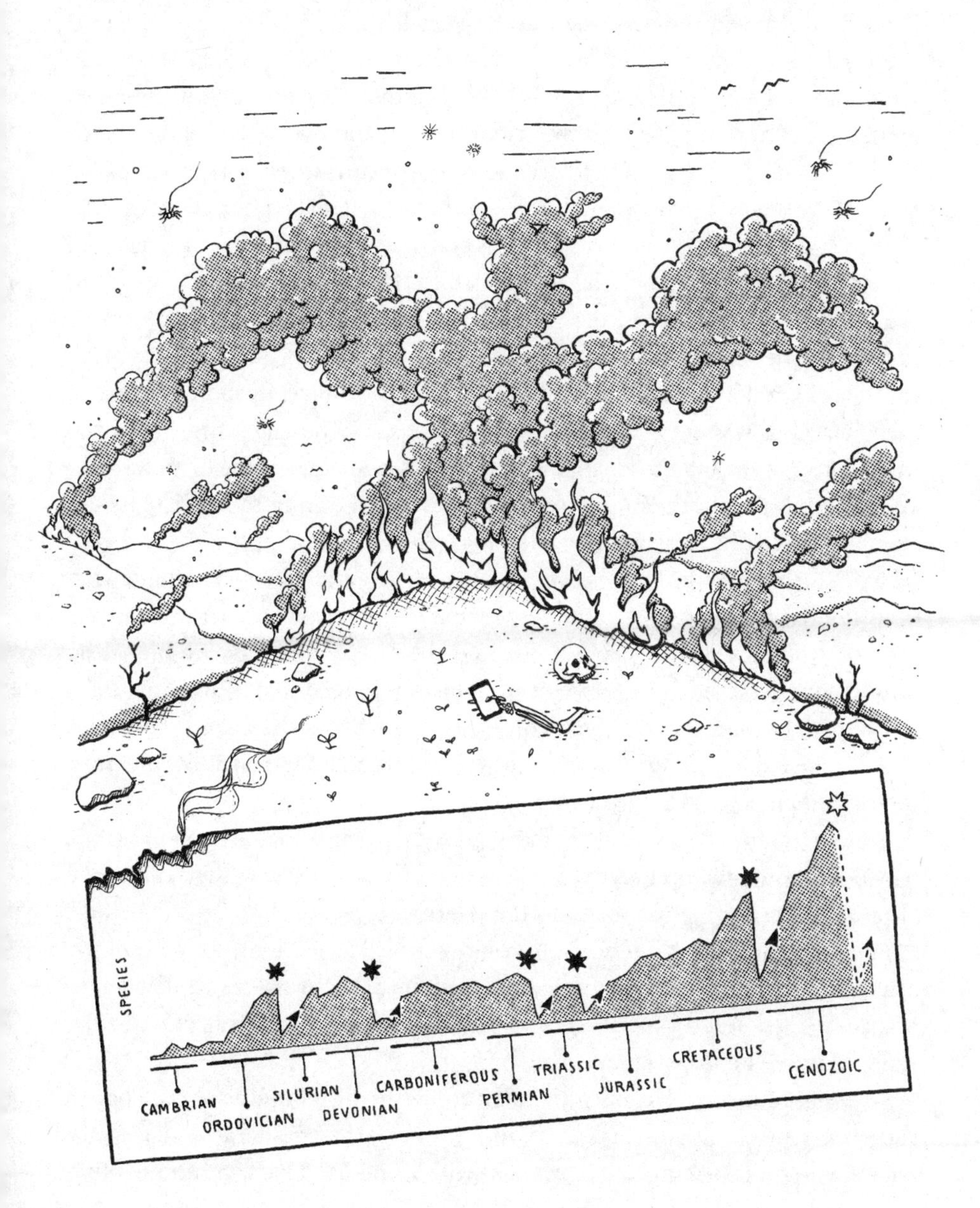

That brings us to the most well known of Earth's mass extinctions: the end-Cretaceous extinction. Roughly 65 million years ago, all of the non-bird dinosaurs ceased to exist. Unlike previous mass extinctions, the smoking gun responsible for the downfall of the dinosaurs is clearly present in the geologic record. A spike in iridium at the end-Cretaceous boundary gives clues to a cataclysmic impact that happened 65 million years ago. Iridium is a metal, rare in the Earth's crust, but common in comets and meteorites. The ten-kilometer-wide object blanketed the Earth with this rare mineral after slamming into the Yucatán Peninsula at more than a thousand miles per hour. The incredible amount of heat generated by the object breaking through the atmosphere would have scorched the Earth for miles around the impact zone. The resulting seismic waves triggered massive tsunamis, which swept across half the world's oceans, making it as far inland as south Texas and northern Mexico. Debris and gases from the exploding asteroid would have greatly changed the Earth's climate for the years that followed. On top of all that, volcanoes of the Deccan Traps in India had been erupting for thousands of years prior. Gases emitted from the volcanoes would have produced caustic acid rain and likely led to changes in global climate. In short, the end of the Cretaceous was not a good time to be a dinosaur. Thus, after more than 160 million years of ruling the Earth, the dinosaurs yielded the throne to mammals.

Assuming the Earth is not completely destroyed during the next mass extinction, we can be sure that life will go on. The ecological process of recovery from disturbance is known as succession, and often proceeds in a somewhat predictable and orderly way. Our best examples of succession after catastrophe are well-studied volcanic eruptions. In 1883 the island of Krakatoa in Indonesia erupted in one of the most impressive explosions in recorded history, unleashing deadly tsunamis and almost completely destroying the island. However, within a year, life was already returning to the smoldering ruins. Eventually a new island emerged from the ocean.

Among its first colonists were ferns and other plants that arrived by wind-borne spores or floating seeds. Spiders arrived early, because they can move long distances by "ballooning" on the wind as baby spiders on silk threads. Larger colonists, including birds, fruit bats, as well as pythons, were observed arriving over the years. And within only 40 years of the eruption, Krakatoa had luxurious forests and over 30 species of birds, which is probably close to the number found on the original island.

In 1980, Mount St. Helens provided another living laboratory of ecological succession in Washington State, exploding in a fury that obliterated 600 square kilometers of forest, scorching Spirit Lake with white-hot lahars and annihilating all life within a 15-square-kilometer blast zone. Scientists arrived as soon as it was safe and found that, within hours, bacteria and algae had recolonized Spirit Lake. The biggest surprise was the contribution of survivors to the regeneration process. Salamanders hibernating deep underground reemerged and quickly colonized otherwise lifeless ponds. Pocket gophers survived surprisingly close to the blast zone, accelerating succession by bringing up nutrient-rich soil and seeds from below the ash layer to the surface. Forest regeneration is now well on its way, with small trees, wildflowers, mule deer, and elk common throughout the blast zone.

While it is sad that you will probably not survive the next mass extinction, take solace in the fact that spiders and gophers will. So each time our pale-blue dot has faced an extinction event, life has somehow managed to bounce back. Eventually we will encounter a future that will include increasing solar radiation and temperatures on Earth rising well over 100 degrees Celsius, thus making it completely uninhabitable. But that is not going to happen in the next 50 million years, so we still have some time.

References

Adams, Douglas. *The Restaurant at the End of the Universe* (London: Pan Books, 1980).

Betts, H. C., et al. "Integrated Genomic and Fossil Evidence Illuminates Life's Early Evolution and Eukaryote Origin." *Nature Ecology and Evolution* 2 (2018): 1556–62.

Melott, Adrian L., et al. "Late Ordovician Geographic Patterns of Extinction Compared with Simulations of Astrophysical Ionizing Radiation Damage." *Paleobiology*, 35, no. 3 (2009): 311.

Alvarez, Walter. *T. rex and the Crater of Doom* (Princeton: Princeton University Press, 2013).

Range, Molly M., et al. "The Chicxulub Impact Produced a Powerful Global Tsunami." *AGU Advances* 3, no. 5 (2022).

Nash, Steve. "Making Sense of Mount St. Helens." *BioScience* 60, no. 8 (2010): 571–75.

Thornton, I. W., et al. "Colonization of the Krakatau Islands by Vertebrates: Equilibrium, Succession, and Possible Delayed Extinction." *Proceedings of the National Academy of Sciences* 85, no. 2 (1988): 515–18.

Whittaker, R. J., M. B. Bush, and K. Richards. "Plant Recolonization and Vegetation Succession on the Krakatau Islands, Indonesia." *Ecological Monographs* 59, no. 2 (1989): 59–123.

Anirban Bhattacharjee (astronomer), Thomas Shiller (geologist), and Sean Graham (biologist) began their science partnership in 2017 while colleagues at Sul Ross State University in Alpine, Texas. Following a beer-fueled discussion at their local dive, it was decided that the three friends should combine their science powers and give a tag-team Nerd Nite presentation. So, what nerdy topic involves astronomy, geology, and biology? The answer: mass extinction!

NERD NITE DURHAM

HOW AND WHY CANCER HAPPENS; or, If You Live Long Enough You're Going to Die

by **Dr. Kerry P. Donny-Clark**

Cancer has been the second leading cause of death in the US since 1933 (take that, pneumonia!), with only heart disease killing more people.[1] In fact, in modern times cancer has even been gaining on heart disease. But what is cancer and why do we get it? Why do old people get cancer much more than young people? Why is cancer so hard to treat?

Cancer is a disease where some cells grow out of control, dividing and expanding across your body. The cancerous cells break down the balance your body needs to function, and eventually the disruption is so great that you die. But what makes those cells go berserk?

Cancer is caused by mutations to your DNA. Almost everyone starts out as a single-celled embryo with nice, happy DNA that encodes all the functional proteins your cells need to grow, divide, and specialize. But as you grow and age, mutagens are constantly attacking the DNA in your cells. Mutagens sound like something out of a comic book, but alas, these mutagens will not give you any powers. Mutagens are things like UV rays from the sun, chemicals from tobacco smoke, radiation, food additives, and more. And even if you live in the dark, eat only kale and blueberries, and shun all smokers, you still aren't safe from them, as chemical reactions in your cells will generate by-products that damage your DNA anyway.

1 https://www.cdc.gov/nchs/data/dvs/lead1900_98.pdf.

With all these mutagens, why doesn't everyone have cancer already? Partly this is because you have a lot of DNA, and DNA itself is somewhat error-tolerant. Your cells also have numerous DNA repair systems that find mutated DNA and fix it. In addition, your cells have systems that cause them to commit suicide (a process called apoptosis) if they have too many errors. This process prevents those cells from becoming cancerous.

To get cancer, one particular cell has to get just the right errors to break all of these protective systems. Remember that DNA controls nearly everything the cell does. For example, a cell's DNA repair system can be mutated so that it can't repair DNA correctly. To become cancerous, the cell has to accumulate just the right set of mutations in various anti-cancer systems. It's like collecting Pokémon, only in this case the mutations gang up and kill you once you collect 'em all.

Mutations are randomly spread across DNA, so just by probability it usually takes a long time for one cell to get the right set of mutations. This is why we see low cancer rates until around age forty, then a steady increase with age.[2] The other side of the coin is that the more time passes, the more mutations all cells have. This means that even if you totally eradicate a given cancer success-

2 https://www.cancer.gov/about-cancer/causes-prevention/risk/age.

fully, we could still expect a different cell to become cancerous in the future. Even worse, we would expect the interval between cancers to decrease as you get older. Thus, the more successful we are at prolonging life and treating cardiovascular disease and other causes of death, the more cancer will dominate mortality. Essentially, if you live long enough, you're going to die.

Or are you?

This was a given when I finished my PhD specializing in DNA damage and repair in 2010. Back then I always said that the only way we could really, truly beat cancer would be to revert mutated DNA back to its original code. At the time, that kind of DNA editing seemed like science fiction, but more recently huge technological strides have been made in the ability to do targeted, specific edits to DNA. This means that scientists could edit DNA just as Matt and Chris edited this contribution, doing a "spell check" to replace cancerous DNA sequences with correct ones. And today, researchers are investigating how gene editing can revert DNA damage to cure or prevent cancer. The technology keeps getting better, and the first clinical trials are showing promise.

So now, maybe if you live long enough . . . you're going to *live*.

Dr. Kerry Donny-Clark is a manager of the Apache Beam team at Google. Before this he has been a professional yo-yo player, an English teacher in Japan, a cancer researcher, an elven fighter/mage, a janitor, a circus performer, and various kinds of software engineer. Kerry likes to build things, from furniture to applications, and his hobby is collecting other hobbies. He lives in the woods of Pittsboro, North Carolina, with his wife, six kids, two cats, and a dog named Krypto.

NERD NITE HOUSTON

ALGAE APOCALYPSE: The Most Important Slime

by **Lewis Weil and Rose-Anne Meissner, PhD**

Us humans, justifiably so, are pretty full of ourselves, as we feel quite confident that we're the apex organism. Over the billions of years of life on Earth, we definitely think we're the most important life-form. After all, we did invent Tater Tots, smartphones, viral cat videos, and global warming. We're a pretty big deal. But algae, particularly the cyanobacteria, a photosynthetic bacteria, were and continue to be the most impactful organisms, and have done more to shape life on Earth than humans ever could. Cyanobacteria have been around for billions of years, are responsible for life on Earth as we know it, and sorta caused the biggest extinction events in Earth's history. Twice.

Cyanobacteria were not the first life on Earth, not by a long shot, but they were the first organisms to leave a fossil record. They left behind fossils that are known as stromatolites, which are formed of layers of sediment and mats of cyanobacteria. Sediment settled on the mats of cyanobacteria, then a new layer of the algae grew on the sediment. In fact, the earliest known stromatolite formed 3.45 billion years ago! Remarkably, there are living stromatolites on Earth right now in the freshwater cenotes of Cuatro Ciénegas in Coahuila, Mexico, and in Shark Bay in Australia. Modern cyanobacteria are the distant descendants of those first fossils.

Apocalypse #1: A couple billion years ago, cyanobacteria evolved a neat trick, a process of using light from the sun to gather energy and produce oxygen as a by-product. Molecular oxygen was a waste product—gas trash. At the time, there

WORLD'S MOST INFLUENTIAL ORGANISMS

was no oxygen in the atmosphere, and life had evolved under these oxygen-free conditions. For millions of years, the oxygen produced by cyanobacteria reacted with metals by *oxidizing* them to produce rust. In fact, there's a layer of rust around the entire globe left by rust settling out of the oceans and lakes. Once all the exposed metals rusted, oxygen at last started building up the atmosphere. Whereas this oxygen buildup is critical for those of us who breathe oxygen today, it killed most of the life on the planet at that time. You see, oxygen was toxic to the organisms that had evolved without it.

Cyanobacteria are patient; they're in no hurry. They waited a couple billion years to strike again.

Apocalypse #2: In the Eocene epoch, about 50 million years ago, the Arctic Ocean (North Pole) was surrounded by land on all sides. Much like the Great Salt Lake in Utah today, fresh water from rivers would flow into the Arctic, bringing in minerals and salts. But there were no outlets from the Arctic except for evaporation. Thus, these salts built up in the Arctic, making it saltier than the regular ocean, with a thin layer of fresh water sitting on top of dense saline. At this time, there was also way more carbon dioxide (CO_2) in the air than there is today. Though humans have increased the concentration of CO_2 by about 50 percent since before industrialization, causing a little problem known as global warming that also might kill us all, during the Eocene, CO_2 was about *1600 percent* higher, at approximately 4,000 parts per million (ppm). Earth was ripping hot, and the Arctic was tropical. The North Pole wasn't frozen, but it still went through summers when the sun never went down, and early-winter days were enveloped in complete darkness.

In the Arctic Ocean at the time, there was a tiny floating fern called Azolla growing in the thin layer of fresh water. Amazingly, Azolla are still around today. This tiny plant is special because it is host to a symbiotic cyanobacteria. Besides the ability to make sugars from atmospheric gas, some cyanobacteria can grab nitrogen out of the atmosphere and use it to make amino acids for proteins. This superpower means some cyanobacteria, including those associated with Azolla plants, can get all their food from thin air. By sharing this food with their hosts, cyanobacteria allowed Azolla to thrive in places where other plants cannot, such as the Arctic Ocean during the Eocene.

During Arctic summer, the Azolla ferns would grow continuously in the ever-present sunlight, capturing carbon through photosynthesis. During Arctic winter, they would die and fall into the deep hypersaline waters in which they would be preserved, sequestering the carbon they had absorbed. This process continued for 800,000 years. In a process akin to climate change in reverse, the sequestered carbon cooled the surface of the seas to 72 degrees Fahrenheit, which allowed plant life to explode on land, further sequestering carbon. The drop in temperature helped some life on Earth but, again, killed most of it.

This Cyanobacteria Apocalypse Part II created the era in which we now live, on a relatively cool planet covered with grass, trees, and ice at the polar caps (for now). With a track record like this, one has to wonder: What are cyanobacteria going to do next?!

Lewis is a scientist turned financial planner and ecological restoration practitioner. He is no stranger to the Nerd Nite stage, being a former boss of Nerd Nite Austin, and has presented on shark sex, apocalyptic algae, and an ode to a beloved giant clam. He has only fallen off the stage once.

Rose-Anne Meissner, PhD, is a science writer, editor, and educator living in Austin, Texas. She and writing partner Lewis Weil are currently working together on a book about financial planning, weed, and the wealth of community.

Space, the Big and the Beautiful

The vastness of space both confuses and scares me. It confuses me because I'm simply unable to comprehend its size. Knowing that our dear planet is a speck within our galaxy, which itself is a speck among uncountable other galaxies, simply makes no sense to me. Only a few times have I consciously made an effort to try to wrap my head around the vastness of it all, but each time I gave up within moments because it was an exercise in futility. I'm a quitter sometimes. And space scares me because it makes me feel like I'm unable to understand a simple concept. If the Big Bang was caused by, at its essence, a giant ball of gas waiting to explode, I feel dumb when I ask, What was one inch off said gas ball's surface? Or what was a billion miles away from it? And then another billion miles away from that? There was something, right, even if just infinite darkness? Something? Ahhhhhhhh-hhh! Shhh . . . it'll be okay. *crawls into fetal position*

But we here at Nerd Nite want to reassure you (read, me) that space can be fun. There's poop in space. There's real gravity but there's also artificial gravity. There are asteroids, maybe tiny life-forms under Europa's icy surface, and some of the most beautiful things that are billions of light-years in size that your naked eye will never see without the assistance of giant tubes and mirrors. And lots of BS, too. So, dear reader, please turn to the next page to boldly go.

—Matt

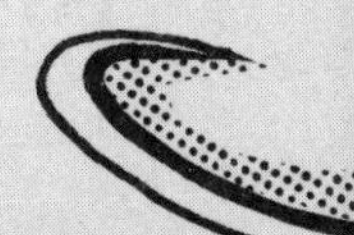

NERD NITE PITTSBURGH

BULLSHIT IN SPACE:
An Astronomical Adventure Through Cosmic Misinformation

by Ralph Crewe

I first became aware of Nerd Nite in 2019 when I was recruited to give a presentation. At the time, I was working at a local planetarium and observatory hosting astronomical events in which I piloted a cutting-edge digital planetarium, flying across the gulf of space. I could transport the whole room to the moon, planets like Jupiter and Saturn, an uncountable sea of stars, brilliant star-forming nebulae, titanic swirling galaxies, and even the very edge of the universe itself. We also had a research-grade telescope on our roof, where I showed thousands of people real-life astronomical objects. Have you ever shown an eight-year-old the rings of Saturn for the first time in their life? The sense of wonder, awe, and curiosity I gave people made this one of the most profoundly fulfilling jobs I've ever done.

Unfortunately, due to various factors, I also spent a fair amount of time talking to people about various astronomical misconceptions, or, as you can see in the title of this contribution above, *bullshit in space.*

Astronomy and cosmology deal with objects and phenomena far beyond most people's tangible experience, and so as people let their imaginations run wild in contemplation of the cosmos, bullshit abounds. I will attempt to touch on a few of the more common flavors of bullshit that I regularly encountered.

One of the most frequent inane topics that filled my time was whether the moon landing was faked. Here are a few of the reasons why that's bullshit:

- If NASA was willing to fake that kind of thing, it 100 percent would have faked a Mars landing by now! While there have been a ton of brilliant space programs since 1969, nothing compares to the grandeur of *Apollo*.

- The USSR would never have let the US get away with such a fake. The fact is that the Soviets never publicly questioned the validity of the moon landing.

- Some folks point out that in photos and film of the lunar surface, we can't see background stars in the blackness of space. These folks usually haven't considered that these images were all captured *during the day*. Sirius, the brightest star seen from Earth, is literally thirteen billion times dimmer than the sun. No camera existed in the 1960s with anywhere near the dynamic range necessary to pick up stars during the day on the lunar surface.

- One of the most compelling pieces of evidence that the moon landing was, in fact, real is the presence of retroreflectors on the surface that were left there by *Apollo* astronauts. Lasers are routinely bounced off these reflectors to measure the moon's orbit (to astounding precision), and that just could not work if the landing was fake.

Perhaps the most blatant steaming pile of astronomical misinformation is the notion that the Earth is flat. I controversially refer to this as "flat-shaming" to explore how absurd this idea is.

Even in antiquity, the Earth was known to be spheroid. In his groundbreaking series *Cosmos*, nerd hall-of-famer Carl Sagan famously explained Eratosthenes's brilliant experiment comparing the angle of shadows in Syene and Alexandria in ancient Egypt. Not only did the famous polymath prove that the Earth is round, but he also measured its circumference within a couple of percentage points!

Of course, it's easy to see that the Earth is a sphere (really, it's an oblate spheroid, but close enough) today. There are mountains of *pictures and videos* from space. If you don't trust that, note that every other large celestial body, from the moon to the sun, to all other planets, are also spheres. The shadow of the Earth during a lunar eclipse is round. Observe a ship at sea from far enough away, and you'll notice that the bottom of the vessel is obscured by the curvature of the Earth. Et cetera.

Also, if there really was some vast conspiracy to trick us into believing the Earth was round, who would benefit? Globe manufacturers? Bullshit!

The last topic I'll mention was the one that was hardest to deliver. While Nerd Nite always draws a delightfully diverse audience, I didn't expect to encounter too many flat-earthers or moon-landing deniers. I knew, however, that I would piss some people off when I called out the zodiac.

Astrology and the zodiac are beloved by millions. People find great joy and satisfaction in reading their horoscopes and identifying as a sassy Sagittarius or whatnot. Unfortunately, it's all bullshit. Not only do the positions of the sun, moon, and planets at the time of your birth have no measurable effect on you, but the zodiac signs don't even line up with the dates used in the newspaper. Those dates are based on the old Julian calendar, which is off by weeks from our current Gregorian calendar.

Even more glaringly, the sun passes through not twelve but *thirteen* constellations every year. In fact, if you were born between November 30 and December 17, your star sign (which, again, is meaningless) is Ophiucus!

Of course, there's plenty more astronomical misinformation out there that I had to deal with during my time in the planetarium. Once you get past that, however, there's also an enormous amount of brilliant science, a deep curiosity, and a wonderful spirit of exploration that has rocketed us into our current understanding of the cosmos. We are truly in a golden age of space exploration, astronomy, and cosmology; no other society has ever come close to the comprehensive understanding of the universe that we are privileged to enjoy. Just think how much more we could discover and explore if we spent a little less time on bullshit!

Ralph Crewe spent ten years working full time at a major science center, where he helped to run astronomical programming for the planetarium and observatory. He works in software now but also is active in the YouTube educational space and, perhaps most notably, is one of the bosses of Nerd Nite Pittsburgh.

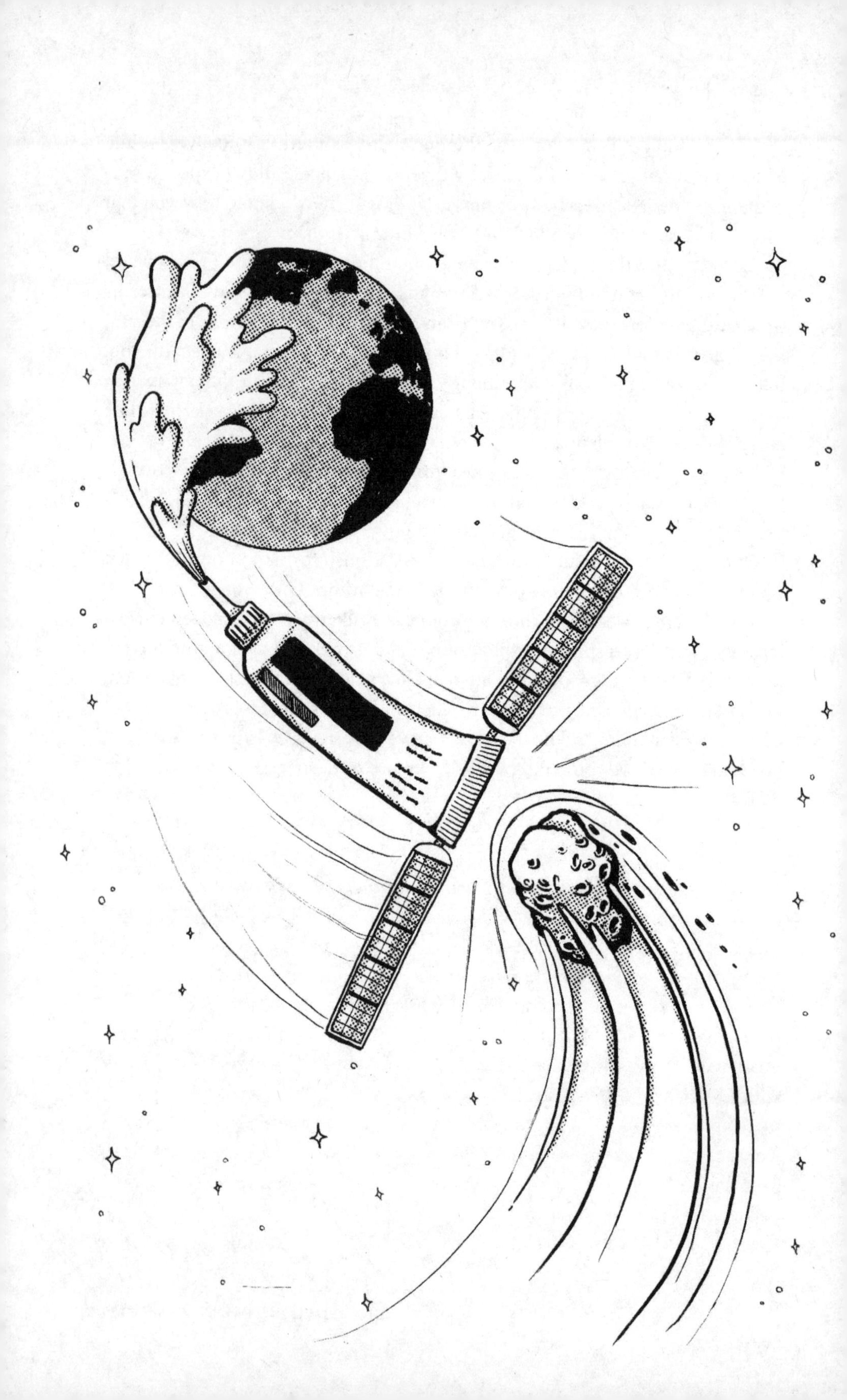

NERD NITE ORLANDO

PREPARATION A:
Our Relief Against Severe ASSteroids

by **Dr. Derek Demeter**

Are you concerned about planet Earth being on the path of a catastrophic asteroid impact that could wipe out all life as we know it? Could this be the one that causes the next extinction event like the one seen during the end of the Cretaceous period sixty-six million years ago? Will it happen in our life-time? Well, worry no more—Preparation A is here for you!

Our solar system contains millions upon millions of small bodies called minor planets that orbit the sun; these include asteroids and comets. Over time, many of these objects collide with each other, resulting in debris that disperses throughout the solar system. Some of this material may orbit the sun in such a way that it passes through our planet's orbital path. Over twenty-nine thousand of these objects called near-Earth asteroids have been documented. Using observations made from observatories around the world, NASA's Center for Near Earth Object Studies models and characterizes the paths of these potential threats in much the same way as we track the path of hurricanes or other weather events on Earth. Thankfully most of them do not cross paths with Earth too closely and pose no threat, but a small fraction of them do get too close for comfort (approximately four and a half million miles or closer), which are described as potentially hazardous asteroids. And size does matter in this case, so a hazardous asteroid must be about 460 feet or larger to cause catastrophic damage to Earth; the Cretaceous extinction event was thought to have occurred from an impact by an asteroid the size of a few miles in diameter.

So how likely is it that we'll need to apply Preparation A to our planet? Well, there is a small chance (about 2.7 percent) that in 2029, an asteroid called 99942 Apophis could impact the Earth. Although that may seem like a low probability, it is the highest we have seen in a long time. So, what can be done? Should we blow it up like we see in the movies? It turns out that Preparation A is much better administered when done via a gentler and more calculated approach.

In October 2022, the Double Asteroid Redirection Test, also known as DART, successfully slammed into the asteroid Dimorphos. After the impact, Dimorphos had a slightly slower orbital speed, which altered its orbit around its parent body, Didymos. Further studies must be done, but this test did prove that we can in fact alter the orbit of an asteroid and deflect it from impacting Earth. Future missions like the European Space Agency's Hera mission will continue this research and provide us with an even better plan for the inevitable future.

Some asteroids also contain an abundance of very valuable rare-earth minerals such as platinum, silver, and gold. Thus, us Earthlings are certainly interested in retrieving such asteroids to bring them back to Earth to mine. Imagine gold, an extraordinarily precious and valuable resource here on Earth, becoming as abundant as sand.

The human species has one very powerful tool that no other animal on our planet possesses, our brain. With our determination and ingenuity, we can apply Preparation A in a way that the fear of an asteroid flare-up will be a thing of the past.

Derek Demeter began working at the Emil Buehler Planetarium at Seminole State College of Florida in 2003. In 2007, he became the planetarium director and immediately began the process of writing and producing shows for the planetarium. Since then, his passion for teaching people about the wonders of the universe has earned him many accolades and worldwide recognition. Derek recently served as president of the Southeastern Planetarium Association. He enjoys promoting science beyond the planetarium with his work as an astrophotographer. He has been featured in NASA's Astronomy Picture of the Day, Astronomy Magazine, *and other publications.*

NERD NITE RENO

LIFE UNDER THE ICE OF EUROPA

by Guillermo Garcia Costoya

When we think about what life might look like on other planets, our imagination is kind of trash. We always picture organisms looking like what we know, evolving on planets like Earth, and our preconceptions skew how we search for life in the universe. We look for exoplanets on distant stars that just so happen to be close enough to their suns not to freeze but not close enough to get roasted. Now, that's sad, isn't it? We are not going to get close to any of those stars and planets anytime soon, so our only hope of meeting aliens is for them to come to us and try to destroy us *Independence Day* style. But what if I told you that we might not have to look that far? What if I told you that there might be life here, in our own solar system? Remember the planet Jupiter? The big one; the one made out of gas.

Well, Jupiter has a bazillion moons, most of them small, but there are four pretty big ones that we call the Galilean moons: Io, Calisto, Ganymede, and Europa. The first three are cool and all, but Europa is where the good stuff happens. Imagine a literal ball of water ice a bit smaller than our own moon; though it may not seem too interesting at first, it turns out that since 2014 we've been seeing huge plumes of water vapor coming out of it. Yes, vapor plumes, like the geysers in Yellowstone, but tall enough that they are visible from Earth through a telescope! Nuts, I know. And if you can add two plus two together, you know where this is going.

If there is water vapor coming out, the water ice must become liquid at some point to get to gas, right? Here's the best explanation we have so far: Jupiter, all massive and all, has an insane gravitational pull, so big in fact that it literally squeezes the orbiting Europa like a stress-relief ball. All of that squeezing, then, might turn into heat at Europa's core, and that heat melts the ice, and BOOM!, you get an ocean.

As a matter of fact, some scientists speculate that there might be an ocean of liquid water up to a hundred kilometers deep under a thick-ass layer of ice. To put that into context, that would be more water than in all of Earth's oceans combined. From life around here we know two things: First, where there's water, there's life; and, second, life can happen at the bottom of the ocean. Hydrothermal vents on Earth, for example, are areas at the bottom of the ocean, between tectonic plates or near volcanoes, where the heat from Earth's core is released in huge plumes of water vapor and other gases. Some scientists predict that the bottom of Europa's oceans might be filled with similar areas and that these might be the cause of the plumes we see.

On Earth, these areas support immensely rich ecosystems that can function with absolutely no sunlight, so why couldn't life happen on Europa's hydrothermal vents as well? Imagine all the weird creatures that might inhabit Europa's oceans. Most likely, our little brains don't even have the mental capacity to picture how different life might be over there. So when will we see it? I mean . . . NASA and the European Space Agency have been saying that they are going to send the *Europa Clipper* mission over there in 2023, then in 2024, then in 2027, then "sometime in the 2020s." So no, we are not close, but at least we are closer than planets on other stars!

For now, our only option is to patiently wait and try to imagine what life, evolving for millions of years on a planet with no light, might be like. Regardless of the outcome, whenever we see it, it might be one of the most important moments in history.

Originally from Spain, Guillermo is a graduate student at the University of Nevada–Reno studying how climate change will impact lizards (spoiler alert: not great).

NERD NITE LOS ANGELES

ARTIFICIAL GRAVITY IN SCIENCE FICTION

by **Erin Macdonald, PhD**

In order to talk about how gravity works in science fiction, first we need to understand how gravity works. Some of you may have learned the Newtonian rules of gravity: that it is a force between two objects that decreases based on how far apart the objects are. However, in the late 1800s this theory began to fall apart. See, as Mercury (the closest planet to the sun) orbits, it "precesses," which means the orbit itself shifts, tracing out a flower shape. If we assume the rules of Newtonian gravity, in order for Mercury to orbit in this way there would have to be *another* planet closer to the sun that was tugging on Mercury. For a time this theoretical planet was known as Vulcan! As astronomy improved, it became clear that this planet did not exist. Scientists knew where it *should* be in order to explain the orbit pattern, but it just wasn't there. This is the fun part of astrophysics: It's an interplay between science and observation, where scientists just have to take the different puzzle pieces the universe gives them, build theories based on them, and check future observations to make sure that the theories make sense. Without Vulcan, scientists needed a different way to explain Mercury's orbit.

At the same time as simplistic Newtonian gravity was beginning to crumble, mathematicians were toying with the idea that our universe could be thought of as a four-dimensional fabric: three dimensions of space and one of time. In the early 1900s, Albert Einstein took this concept one step further and proposed that gravity was actually due to the shape of this fabric when mass is introduced.

In other words, it's like putting a bowling ball on a trampoline. The mass (the bowling ball) bends spacetime (the trampoline), creating a "gravity well" that tells other mass how to move around it. If you then flick a marble around the bowling ball, this pattern of movement will approximate a planetary orbit, and the movement of the marble toward the bowling ball is akin to gravity (*Editors' note: it might be helpful to watch YouTube videos of this experiment*). This theory was checked against Mercury's orbit, and lo and behold the orbit was able to precess without another mass there! Further evidence has proved this concept, called general relativity, including using light bending around gravity wells as it passes by (gravitational lensing) and ripples in the fabric due to massive objects colliding (gravitational waves).

Rotational artificial gravity

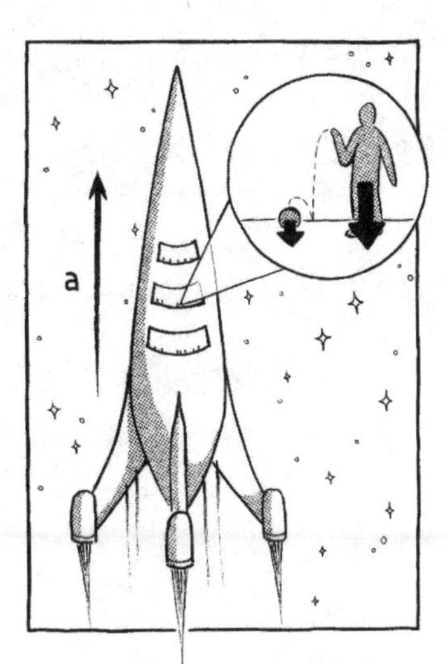

Linear artificial gravity

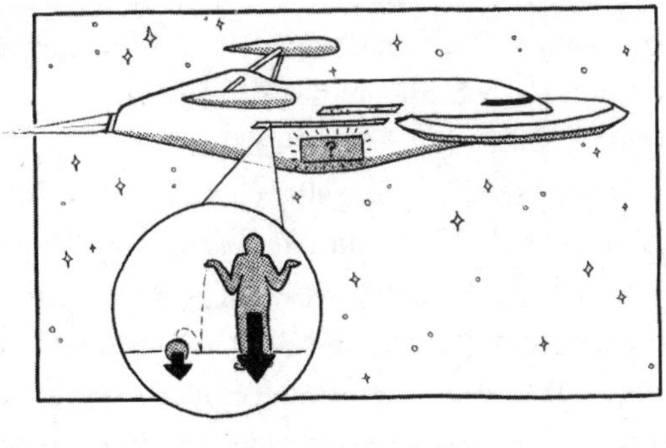

Gravity generator
(just trust us)

When we look at space exploration, it's important to consider the fact that there is no gravity well holding our bodies upright in one direction. Notably, the weightlessness of astronauts in the International Space Station (fun fact: actually due to the fact they are "falling" around the Earth instead of being fully outside the gravity well) has been shown to have negative impacts on human physiology such as loss of bone density and muscle mass. Thus, if we want to consider long-term space travel, we need to actually think about how we could have some form of artificial gravity. This is where science fiction comes in! Because not only is science fiction set in the future or sometimes even in another version of technology but also it is easier to film a movie or TV show or to program a game when we can have our characters just walk around as they would here on Earth.

The most popular and easily recognizable form of artificial gravity in science fiction is rotational artificial gravity. This was first popularized in *2001: A Space Odyssey* and further seen in works such as *The Martian*, Larry Niven's *Ringworld*, and the Citadel's Promenade in the video game series *Mass Effect*. Rotational artificial gravity uses centrifugal force, the force generated while spinning around an axis, to simulate a pull on bodies that can be comparable to gravity. If we are being spun in a circle, then our bodies are always fighting that circular motion and we find ourselves feeling "pulled" to the outside. This is also how they train fighter pilots and astronauts to withstand higher gravitational forces: literally putting them in a centrifuge and spinning at higher speeds. Unfortunately, we don't have these rotational space stations yet because of a combination of cost and size. If you want to do it for a reasonable budget, then your station is going to be small, and a small, rapidly spinning space station will generate pretty nasty vertigo if you're trying to walk around. Thus, it needs to be bigger, which at this stage is prohibitively expensive.

Another way we can simulate gravity is through linear artificial gravity. This uses Einstein's Equivalence Principle, which states that someone in a box accelerating through space at 9.8 meters per second^2 (m/s^2) versus a box sitting on the surface of the Earth feels the same gravitational force. This is used to a great extent in *The Expanse*, where the ships accelerate at 9.8 m/s^2, continuing to gain speed until they are halfway to their destination. They then flip around and decelerate at the same rate. During the flip, the crew only feels two times Earth's gravity, which is totally sustainable for humans. Of course, if you were to keep accelerating at this rate, you would eventually hit the speed of light, but that would be at the very edge of our solar system, which is where the story is confined. The reason we are not able to have this form of artificial gravity yet is that we have not found an efficient enough fuel source to sustain that rate of acceleration.

Finally, there are plenty of other methods of artificial gravity in science fiction, most of which use a "gravity generator" with no further explanation given. The video games *Mass Effect* and *Half Life* do implement some references to literally bending spacetime to create a gravity well at a local level. So while we're not quite able to have artificial gravity in space yet, science fiction has come up with lots of fun alternatives that are absolutely rooted in science.

Erin Macdonald (PhD, Astrophysics) is a tattooed one-woman STEM career panel, who received recognition as a researcher, speaker, engineer, and consultant before undertaking her current career. She lives in Los Angeles and works as a writer and producer. She is also currently the science adviser for the entire Star Trek *franchise.*

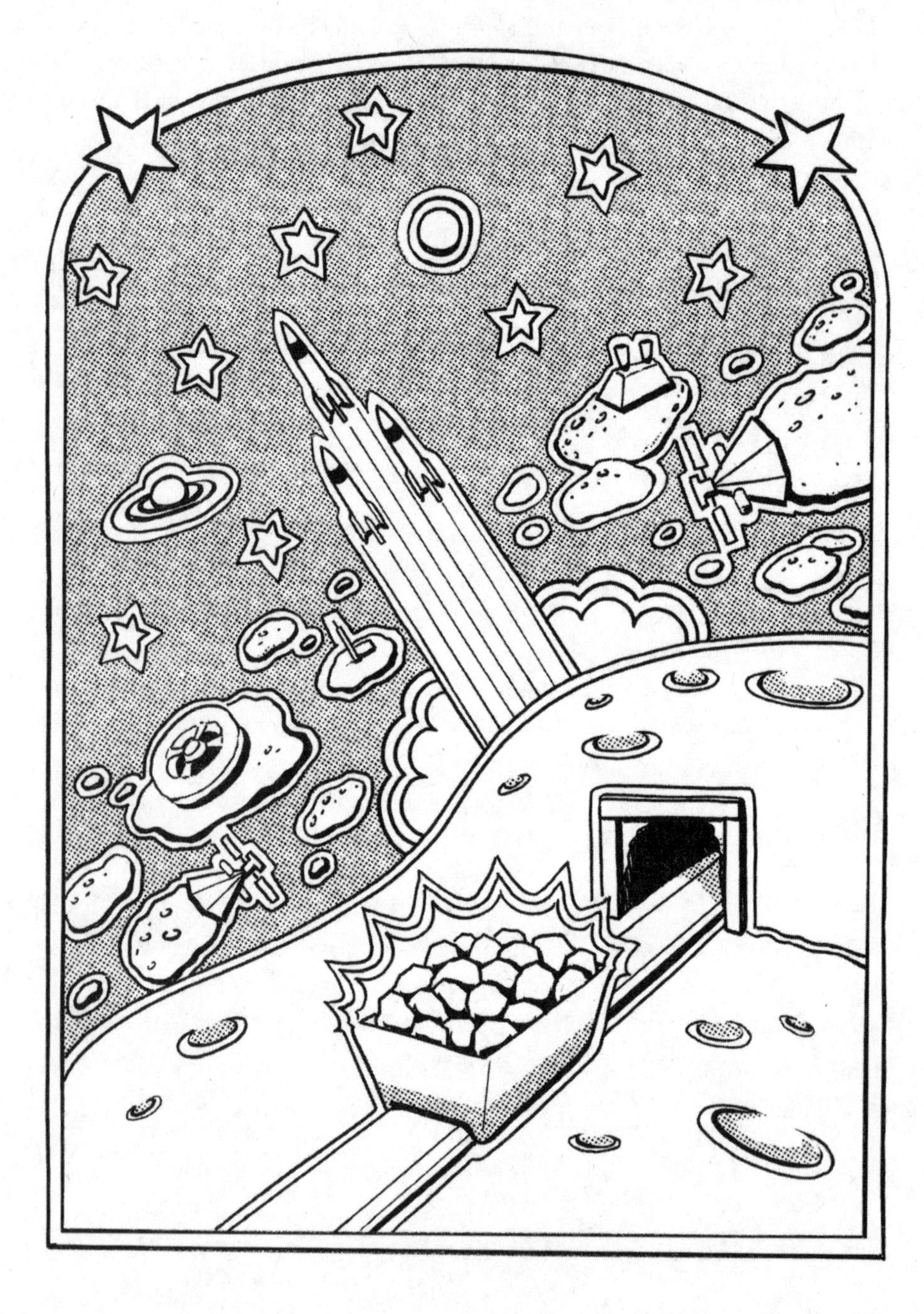

NERD NITE ORLANDO

SKY ROCKETS IN FLIGHT, ASTEROIDS DELIGHT: Asteroid Mining for Science, Profit, and Fun!

by Dr. Zoe Landsman

Headlines about asteroid mining have excited the public by evoking multi-quadrillion-dollar payouts and a futuristic, space-based gold rush. While assigning monetary values to space rocks is click-provoking, it is also economically dubious, and the real story of asteroid mining is even more exciting. Asteroid mining is a crucial step toward humanity becoming a truly spacefaring species. *Spacefaring* means enabling human exploration and long-term presence in space far beyond our Earth-orbiting stations and brief jaunts to the moon. And spacefaring may be more important than just satisfying our itch to explore, should the Earth become uninhabitable due to, say, catastrophic asteroid impact or unrestricted exploitation of our natural resources.

For the uninitiated, asteroids are the roughly four-billion-year-old remnants of planet building. They are the stuff that didn't become Mercury, Venus, Earth, or Mars. Asteroids are mostly made of rock, iron, and nickel, with smaller amounts of other stuff, including water, salts, organic molecules, platinum, and diamonds. Most reside between the orbits of Mars and Jupiter, the famous Asteroid Belt, but a smaller population of asteroids lives closer to home, the so-called near-Earth asteroids—including those pesky potentially hazardous ones.

I've studied asteroids because of their scientific value. Since they are the cosmic leftovers of the formation of the solar system, they hold valuable information about the conditions under which the planets formed, and perhaps tell us more about how other planetary systems form and evolve, too. But what do asteroids have to do with launching humanity into our *Star Trek* future as a spacefaring species? It comes down to the physics described by the rocket equation. (Unfortunately, it always comes down to physics.)

The rocket equation tells us that as you add more mass to a rocket's payload, you consequently need more propellant to achieve a given change in velocity (delta-v)—for example, the rocket needs to reach Earth's escape velocity to go into orbit. But propellant itself is heavy, and so you need more propellant to launch the extra propellant. As you add more and more payload to your rocket, and need more and more propellant to launch the payload, and more and more propellant to launch the extra propellant, you quickly run into a point of diminishing returns when it comes to delta-v. This is a problem for doing really cool stuff in space—building enormous human habitats, for example, or building massive ships that can take humans far beyond the Earth-moon system.

Fortunately, it is possible to overcome this dilemma, to some extent, by using an asteroid as a launchpad. A small asteroid's escape velocity is much lower than the Earth's, so that helps you in your rocket equation budget. You don't need much delta-v to launch from the surface of an asteroid, but you still must get your heavy stuff into space in the first place. This is where asteroid mining is crucial!

You can use the materials in the asteroid to build everything you need for a hub of space activity—the launchpad, the habitats, the spaceships, the rocket fuel, and more. The technology is closer to fruition than you might imagine, with both government space agencies and private industry channeling money into developing ways to use space resources. Though a bigger challenge may be the international cooperation needed for this endeavor, with a little collaboration we realistically have a shot at sourcing materials beyond those in the ground of our own planet.

Dr. Zoe Landsman is a planetary scientist at the University of Central Florida. She studies the surfaces of asteroids and the moon using remote sensing data and analog materials.

NERD NITE EDMONTON

BETTER THAN NASA: Canada's Sample of an Asteroid; or, the Untold Story of the Tagish Lake Meteorite

by Christopher D. K. Herd, PhD

In the lingo of the study of rocks from space, a fireball is a bright streak of light in the sky, lasting several seconds, caused by a rock hurtling through the Earth's atmosphere. The rock was once a part of an asteroid, but due to a bit of celestial fortune, it crosses the Earth's orbit. This meteoroid (more jargon) has quite a ride as it hits the atmosphere at sixty thousand kilometers an hour; the outside heats up enough to melt it and give off light, and pressure on the rock causes it to break apart into pieces. As new surfaces are exposed, more melting occurs and more light is given off, until either the rock fragments are completely vaporized or they slow down enough that they no longer heat up. Anything that survives these few seconds of hell (for the meteoroid) is called a meteorite; often there are dozens or hundreds of individual meteorite fragments scattered over an area kilometers wide.

Professionals like me who study meteorites are called meteoriticists. This is because the study of meteorites is called meteoritics. We are not meteorologists; in fact, I can't tell you the most basic things about meteorology (other than weather is not climate). The root—*meteor*—is the same because at one point, over 200 years ago, scientists thought that it was impossible for solid rocks to fall through the "firmament" (the solid dome of the sky). By this reasoning, these strange rocks that announce themselves with light and sound not unlike thunder must form within clouds, the way that hailstones do. That idea was overturned in the early 1800s when the weight of evidence presented by Ernst

Chladni and others that meteorites were actually solid rocks from outside the Earth was accepted by the scientific community. But the word stuck and here we are.

At about 8:45 AM on January 18, 2000, a fireball streaked across the sky over northwestern Canada. It was observed by hundreds, if not thousands, of people across Alberta, British Columbia, Yukon, and Alaska. Some thought it was "the Americans" testing weapons. Others thought it was aliens—for one fellow from near the small community of Tagish in northern BC, it resulted in a vision of himself as Elvis. He promptly dyed his hair, started wearing rhinestones, and apparently sings his way through Elvis's repertoire on a daily basis. For David Hamelin and Neil Macdonald, it would inspire a short horror film called *Fragments*. For meteoriticists, this was the event of a lifetime.

A few days later, a local resident was driving his truck along the frozen surface of the Taku Arm of Tagish Lake near the Yukon–British Columbia border when he noticed black objects on top of the snow and ice. Instinct told him they were wolf droppings but he had spoken with experts after the fireball and so he stopped to check. Sure enough, these were fragments of what would later be called the Tagish Lake meteorite. On the advice of those same scientists, he picked up the fragments—about four dozen in all—without touching them directly with his hands, took them home, and put them in his freezer. Then it snowed. Any other pieces would have to wait until the spring to be collected.

The first studies of the Tagish Lake meteorite were done by researchers at the NASA Johnson Space Center (JSC), that same facility with all those Apollo samples of the moon. The biggest meteorite fragments were sent, still frozen, to JSC, and pieces of at least two of the fragments were used to try to classify the meteorite. Amazingly, the Tagish Lake meteorite doesn't fall into any known categories. Even more remarkable is that the meteorite contains organic matter from the *beginning of our solar system*! In this context, *organic* doesn't mean life, but it does mean molecules that are important for life as we know it—things like amino and fatty acids, and sugars (PSA: meteorites have no nutritional value). The fact that the meteorite fell on the frozen surface of a lake in the middle of winter in northern British Columbia, and that the finder picked up the meteorites without touching them, and that they have remained frozen ever since, means that these meteorites are possibly the most pristine ever collected.

Where was I in all this? Until 2001, still in grad school. From 2001 to 2003 I was a postdoctoral researcher at the Lunar and Planetary Institute in Houston, right next door to JSC, but relatively uninterested in Tagish Lake and similar meteorites. In 2003 I started as a professor at the University of Alberta and became the curator of the meteorite collection. Throughout this time, though, my father, then curator of the National Meteorite Collection of Canada, recognized the potential of the Tagish Lake meteorite for Canadian meteorite science and became its primary documentarian. Pop's one-hundred-plus-page report was crucial for what would happen next.

Meteorites fall under Canadian Cultural Property law, the same law that applies to works of Canadian art or objects of historical significance to Canada. Five years after their fall, the exported Tagish Lake meteorite fragments at JSC still belonged to the finder, and he wanted to sell them. So he applied to make the export permanent. This opened up the door for Canadian institutions to pay a fair market price for the meteorites in order to keep them in the country. In 2006 I led a consortium that included the University of Alberta (obviously), the

Royal Ontario Museum, Natural Resources Canada, and the Canadian Space Agency. With matching funds from Heritage Canada, we raised a remarkable amount of money in a short period of time and made the deal. I brought the meteorites from JSC to Edmonton in a cooler in its own airplane seat, and the finder brought the samples from his freezer; dataloggers showed that both batches stayed frozen. About three-quarters of the meteorite fragments went into the University of Alberta Meteorite Collection and one-quarter to the Royal Ontario Museum.

In the following years, I did what was expected of me in response to my university shelling out the big bucks to help buy some strange rock from space—I leveraged funding for a specialized lab. At the heart of the lab is a glove box that circulates purified argon gas, inside a walk-in freezer. The lab allows us to break Tagish Lake fragments open in an inert atmosphere and take samples for just about any type of analysis you can imagine. Because the heating during the fireball phase only warms the outer couple of millimeters of the surface of a meteorite, the insides of these fragments have not only never been thawed, they've also never been exposed to the Earth's oxygen or its contaminant-rich atmosphere. No specialized lab had ever been built before, and I can humblebrag that it's even better than anything at NASA. Though in the last couple of years NASA JSC built and commissioned a cold glove box modeled after ours.

The Tagish Lake meteorite is Canada's sample of an asteroid, and it has yielded many secrets so far. It rivals the best samples returned by multibillion-dollar, multiyear space missions by other spacefaring nations at a fraction of the cost. And with it we pioneered an entirely new area of study: the cold curation of meteorites, which informs what humanity will do with future samples from the moon, Mars, asteroids, and even comets.

Chris Herd is a professor in the Department of Earth and Atmospheric Sciences at the University of Alberta. He is the curator of the University of Alberta Meteorite Collection, the largest university-based collection of its kind in Canada. Chris is an expert in the geology of Mars, asteroids, and other places in the solar system from studies of meteorites.

NERD NITE BALTIMORE

THE TELESCOPE THAT BLEW EVERYONE'S MIND . . . PART TWO!

by **Joel D. Green, PhD**

The James Webb Space Telescope, a worldwide collaboration led by NASA but including the European Space Agency (ESA), the Canadian Space Agency (CSA), and many industry partners led by Northrop Grumman, launched on December 25, 2021. And it was absolutely amazing.

Webb launched out of ESA's spaceport in Kourou, French Guiana, after arriving a few months earlier to allow for the testing and stowing of all its parts to be ready for final launch preparation. There were a few hundred points of failure along the way, but they've all been thoroughly tested. The telescope should launch on the *Ariane 5* rocket—the biggest available rocket now that the *Saturn V* is unavailable. It'll go into space and start unfolding like an origami figure. This is because it had to be packed up into a handy shape to fit inside the *Ariane 5*. Eventually it will unfold into a twenty-foot-diameter collection of 18 hexagonal, gold-coated mirrors, sitting on top of a singles-tennis-court-sized sunshield made of tiny Mylar-like strips. This is to keep it cool and protected from contaminating heat sources, since it'll be cryogenically cooled. Since Webb is meant to detect infrared light—which includes heat, as well as a lot of other good stuff—it needs to stay cold. And every bit of this design was engineered to get to that goal.

But I'm an astrophysicist. The real reason I care about Webb is because I want to be one of the first people to *use* it—to make discoveries. This is our generation's Great Observatory. And this telescope is going to change the way we understand the universe. How exciting is it? The best way to understand where we are today is to go all the way back . . . to 1990.

For reference, movies that came out in 1990 include *Home Alone 2*, *Goodfellas*, *Kindergarten Cop*, *Back to the Future III*, and *Teenage Mutant Ninja Turtles*. Even Times Square was still littered with porn shops and was largely in a state of disrepair. Actually, now it doesn't seem that long ago—just a more affordable New York City.

And in April 1990, NASA was preparing to launch the Hubble Space Telescope. And what did people say about it back then? "A vastly sharper, clearer view of the most distant reaches of the universe," wrote Kathy Sawyer of *The Washington Post*.

The popular line was that Hubble would find the most distant galaxies and take pictures of them. So what did NASA do? Release an incredibly boring calibration photo of a few fuzzy stars, showing how they're somewhat less fuzzy with the new telescope. I believe it has learned from this and will not provide such underwhelming images during Webb's six-month calibration period. But it'll be a long time, and the temptation to release news will be great.

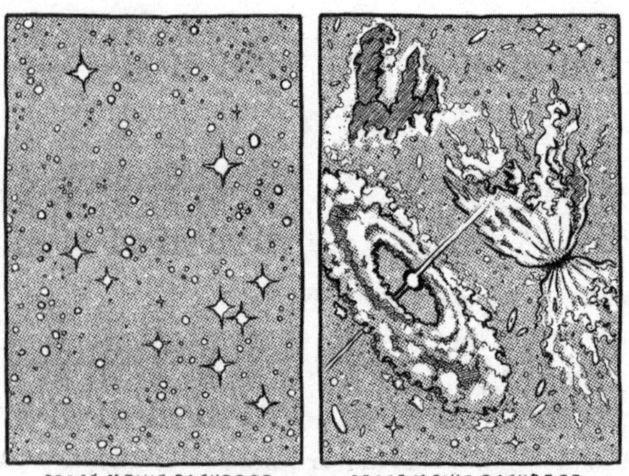

But what did Hubble really see? Take a look at starry backgrounds in movies from the 1970s and '80s. It's not just the lack of computer technology—they are colorless star fields of countless white dots in blackness (think *Star Wars*). Or fuzzy, purple-colored blobs (*Star Trek 2*). Compare that with even the modern *Star Trek* movies, or the *Guardians of the Galaxy* movies—colorful mixes of spirals, clouds, shock fronts, jets, and sweeps of dust and gas, all inspired by Hubble imagery.

My favorite example is T Tauri stars. In *Star Trek: The Next Generation*, there is an episode where they visit a "T Tauri system." Worf explains that these systems are associated with wormholes, and we are shown a brief green sphere-planet-thing. Then the *Enterprise*'s crew is knocked unconscious. I'm pretty sure this whole thing came about because the writers heard this newfangled term in a press release. In the late 1980s, astronomers were really discovering how stars and planets were born, and the actual T Tauri system was a great example of planet formation in progress. But it was all theoretical, or fuzzy infrared imagery. But in 2015, the ALMA observatory took a real image of a disk around a young star—and it is more spectacular than even we had imagined. But Hubble and modern telescopes revealed so much more.

Here are five things astronomers in 1990 believed to be true:

- The universe is expanding, but unsure if it will keep expanding forever, or collapse back on itself. But the expansion is definitely slowing down.
- Other stars have planets like ours—probably.
- Nothing from other solar systems can reach us.
- Stars and planets are born inside nebulae.
- Black holes are a really cool theoretical concept but hard to see.

In 2021, here's how we're doing, thanks to Hubble, ALMA, the Event Horizon Telescope, and others:

- The universe is not only expanding, but accelerating. Also there are ten times more galaxies than we ever thought.
- There are five-thousand-plus known exoplanets, and almost every star has several of them.
- Interstellar asteroids buzz us occasionally. Also Jupiter gets whaled on by comets regularly.
- Stars and planets are born in beautiful knots and pillars within clouds of gas and dust, which make the most inspiring pictures.
- Black holes are real and make good special effects in movies.

So what will Webb see? In 2021, NASA's Eric Smith stated in press releases, "The observatory will detect light from the first generation of galaxies that formed in the early universe after the big bang, and study the atmospheres of nearby exoplanets for habitability."

Well, that sounds a lot like what they said about Hubble, except for the addition about the atmospheres of exoplanets. Progress! But really, as you can see, it's all what we *can't* currently imagine that makes the launch of a Great Observatory so exciting. So who knows?

But because Webb is close to my heart and my work, I can't leave it at that. Here are my top picks and predictions for Webb discoveries:

- Derive the ability to explore and measure the diversity of planets and their atmospheres.
- Understand the composition of the dust and gas that eventually form into planets, connecting to the origins of Earth and the solar system.
- Directly measure the shape of protoplanetary disks.
- See the moments when stars and galaxies began to dominate the early universe.
- Determine the ultimate fate of the dust around dying stars, and the effects of their extreme radiation on their surroundings.
- Track the flares of supermassive black holes, like those in the center of our galaxy.

Someday, our artist renderings of these phenomena will stack up against real pictures, just like the T Tauri system.

Join us for part 3 of this contribution in about 2050!

Dr. Joel Green is an instrument scientist and astronomer at the Space Telescope Science Institute, working with the Hubble and Webb Space Telescopes. He studies the birth and occasionally violent death of planets and planetary systems, in the hope of uncovering our own solar system's past history. Previously, Joel was the project scientist in the Office of Public Outreach, helping bring the science of the Hubble, Webb, and Roman Telescopes to the general public, and has appeared on National Geographic and PBS Nova. *Prior to joining STScI in 2015, he was a research scientist at the University of Texas–Austin. A New Yorker who spent twelve years in Upstate New York, lived for six years in Austin, and is now a resident of scenic Baltimore (and co-boss of Nerd Nite Baltimore from 2016 to 2020), he is fond of crab and barbecue, cowboy boots and hats, sunshine, and Civil War battlefields.*

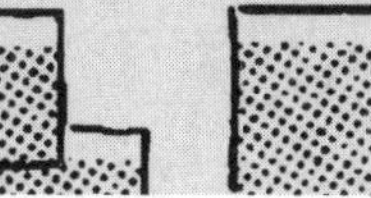

Tech (High and Low)

Technology evolves at a breathtaking pace. It's incredible to consider that any one of your mobile devices contains more computing power than NASA mission control in Houston had at its entire disposal during the moon landing in 1969. Though I suppose it's disillusioning to admit that instead of helping folks to explore our universe, I'm instead more likely to grab my phone, push a few buttons, and find out who played Sasha in *Always Be My Maybe* (Ali Wong), find out how bad traffic is on the New Jersey Turnpike (only a few orange spots approaching the Delaware Memorial Bridge), or, gasp, actually dial—not text—my dad. He's swell and I like hearing his voice. Yet 125 years ago there were no cars, planes, computers, servers, drones, e-bikes, Wonderbras, batting gloves, microwave ovens, Kevlar, atomic bombs, dating apps, or Jell-O (regular or pudding pops). What a time to be alive!

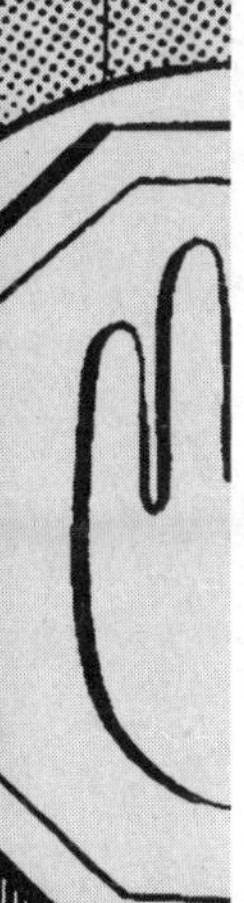

While the term *technology* is basically a catchall for pretty much everything these days, from bits and bytes and hardware, to food, fashion, and transportation, there's still so much to learn and understand. For example, while we all know that the AI robots will either kill or enslave our species by the end of the century, we're not exactly sure how they'll kill us, or, if they put us to work for them, which tasks we'll be assigned. I for one would opt for cuddly human pet, but I'll likely be dead by then so this won't affect me. But what will affect me—and most of you—today are things like the GMOs we consume, the cyborgs we may become, how computers see, how to choose mates via the web, and why nuclear fusion could save us. And guess what? You can read about these last few things . . . now! Just turn to the next page. Easy.

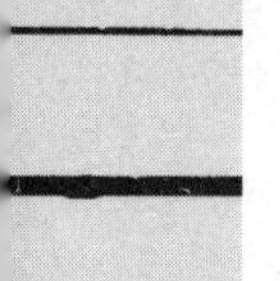

—Matt

WE DON'T UNDERSTAND IT,
THEREFORE IT MUST BURN!

NERD NITE FARGO

"THEY'RE PUTTING ACID IN OUR FOOD!": The Everyman's Guide to Thwarting Fear and Understanding GMOs

by Tracy Kurtz

I was cruising around some interwebs forum a long time ago and some anti-GMO jerkwad had decided to cornerstone his argument with a rant proclaiming that deoxyribonucleic acid couldn't be a real thing, that scientists were just making things up to confuse normal people, because if there was really acid in your cells, the cells would dissolve and you would die. Le sigh.

Acid is a chemistry term for any molecule that can donate/lose a proton to a solution or a chemical reaction (a proton is just a naked hydrogen atom). The more rigorous the donation, the more protons / naked hydrogens in a solution, and eventually your pH is low enough to dissolve secret agents in tanks while you foolishly monologue your villainous plans to them.

But many things, like DNA, are only *weak* acids—sure, they kick out a proton here and there, but it's not nearly enough to be dangerous. The aforementioned jerkwad couldn't tell pH from BO and relied on a clear example of some wisdom I gleaned from watching X-Men cartoons in the 1990s: that people fear what they don't understand.

Well, I'm here to thwart your fear of GMOs by fighting said fear with the best tools we have: understanding and science. We'll back up to what DNA is, how it expresses proteins, and how we use our control of these molecules to engineer better, more sustainable, more nutritious food.

You have probably heard of or studied DNA before, but to be prudent, let's review. DNA is the hereditary material present in the cells of (nearly) all living things—you, your dog, plant cells, bacterial cells—all of it (including all the food you eat, from the stalk of your asparagus to your sous vide rib-eye steak medallions). And it takes on the form of a double helix—you know, it looks like a ladder that has been twisted around itself. The vertical bits of the "ladder" are a sugar phosphate backbone, and the rungs are base pairs of guanine, cytosine, thymine, or adenine (frequently abbreviated G, C, T, or A).

The exact molecular structure of these items can be ignored for our purposes; the important part is the sequence in which these base pairs are strung together—literally the order of your A's, T's, G's, and C's. Each group of three base pairs is called a codon, and each codon codes for a specific amino acid. String multiple amino acids together, and BAM!, you got yourself a protein. And the specific string of DNA bases that codes for a single protein is called a gene.

If DNA is the blueprint of life, then amino acids are the building blocks, and proteins are the larger, more complex building blocks.

GMOs (which is short for "genetically modified organisms") are organisms that have had their genes—aka the blueprints—altered, by either rewriting, adding to, or deleting part of the DNA sequence so that the proteins for which it codes are different.

People have been genetically modifying organisms for thousands of years using other methods—you're probably most familiar with cross-breeding different species to create hybrids (yes, this counts as genetic modification), but you should absolutely Bing "ruby red grapefruit radiation" and learn about plants that were intentionally exposed to radioactive material to randomly mutate their DNA in the hope of getting lucky and finding a delicious mutant. Yes, it's absolutely true, and spoiler alert: It wasn't just grapefruit.

Of course, there are several other more grassroots methods of genetic modification I haven't listed (I'm working on a word limit here), but what they all have in common with hybrids and (especially) radiated mutants is that they are somewhat of a crapshoot—you're hoping for the best with no guarantee you'll get it.

These old-school methods also limit you to genes already present in the species, robbing us of the miracles of golden rice or rainbow papaya, which are transgenic organisms, meaning that the plants contain one or more genes copied from another organism and inserted into the plant's genome using recombinant DNA technology. When people are freaking out about GMOs, they are usually referring specifically to transgenic organisms, even though the term *GM* has much broader applications (see above).

Golden rice contains three additional genes—two from a daffodil and one from a bacterium—which allow it to produce beta-carotene (a precursor of vitamin A), and therefore fight vitamin-A-deficiency-induced blindness.

Rainbow papaya saved the Hawaiian papaya industry from crashing due to infection by the ringspot virus—a single gene from the protein coat of the virus was added to the plant, effectively vaccinating it to the virus.

So why such radical mods? Ultimately, we are racing to feed people. The population, and therefore the nutritional demands of our planet, are expanding like never before. We need to grow more food to feed everyone and we are out of space to expand farmland, so we need to increase the productivity of the land we have by increasing the yields on the crops we produce. Genetic modification can make our food crops stronger by increasing their pest or drought resistance or increasing the crop's yield. It can make food more nutritious, more aesthetically pleasing, or slower to spoil. Recombinant genetics are another tool in our arsenal that allow us to make precision edits to our food instead of blasting it with gamma rays and hoping the resulting monster understands, "Hey big guy, sun's getting real low."

Yes, messing with DNA is freaky and amazing and as close to godlike power as man should ever get, but we've already been doing it for thousands of years with crossbreeding, grafting, and other methods of genetic experimentation. Not to mention that transgenic plants are studied and tested seven ways from Sunday, including multiyear, multilocation evaluations in both greenhouse and field environments for effects of the transgene and overall performance of the plant, including looking for environmental effects and food safety.

Unfortunately, like all scientific miracles, GMs also suffer the side effects of greed and the not-always-well-studied court of public opinion. I won't get into the patent issues surrounding GM seed, and blindness-preventing golden rice was put on hold for years by misguided activists. But can we afford to discard a critical technology over continuously disproved worries while people are starving or sick? No. Should we continue to be thorough and careful with our development of these products? Absolutely. Do we need to worry about acid in our food? No—it's been there all along.

Tracy is just a has-been chemist living in a molecular biologist's world. She enjoys horses, sarcasm, beer, and exactly four other things.

NERD NITE PITTSBURGH

WHAT I LEARNED ABOUT DATING APPS (GENERALLY) AFTER I SPENT FIVE F**KING YEARS STUDYING THEM FOR A PHD

by **Dr. Nicolette Wei Mei Wong**

After spending five years fielding questions by my very Asian parents about what I am trying to do with my life studying dating apps for a PhD—well, I got a doctorate, and this small excerpt to share with the world. Whoop whoop.

My major takeaway? People prefer to be wishy-washy about why they are using dating apps. This is one of the annoying things about being part of a dating app. This is pretty much as annoying as people who only put up group photos in their profile, which results in a confusing and unwanted Where's Waldo situation. But for some bizarre reason, THIS WORKS.

If someone on a dating app asks you, "What are you looking for?" these are the responses that are more likely to help continue the conversation: "I don't know, bored," "seeing what's out there," "I am Sarah's mum and I am helping her see what's out there." But if you responded, "Oh, I'm on dating apps because I'm looking for someone to be my baby daddy to help me birth twenty babies to form a baby army," you are much less likely to be viewed positively and will more than likely be ghosted at the talking stage.

So even if you really are looking for a baby daddy to help you birth twenty babies to form a baby army, it's best to remain ambiguous when your potential baby daddy asks you what you're looking for. Ambiguity is good.

Now, the two primary reasons for this phenomenon are: (1) to not look desperate; and (2) to keep your secondary options open (like placeholders until the "real thing" comes along. It is important to note that "real thing" can mean different things to the dating app user. It could be marriage, a non-monogamous partner, a sex partner, or someone to go to a Brony convention with).

People also often take motivations at face value and fail to realize that people are complex with layered motivations.

For example: If a dating app user says they want a marriage, the recipient tends to immediately assume that the individual wants to tie them down at first moment's notice and is looking to marry the first person that says yes. This is despite the fact that the marriage-seeker is more likely to mean this: I am looking for a partner that has the qualities fit for marriage and may have marriage in mind if there is indeed compatibility. However, I am also open to exploring different paths with you according to my needs and desires at the time. For example, I hear a baby army is cool.

In short, through an accidental design of our own doing, we have created a dating app environment that punishes honesty, however tactful. So the unfortunate moral of the story is: When someone asks you why you are on a dating app, lie your ass off.

Nicolette Wei Mei Wong has a PhD in Anthropology and focuses on being a nosy asshole who asks people about their dating app habits all the time. When she is not interrogating random folks about why they upload pictures of themselves holding a fish on dating apps, she is seen giving TEDx Talks on UX research. She is currently a UX researcher in Big Tech.

NERD NITE TORONTO

ADVENTURES IN HUMAN-POWERED FLIGHT

by David Donaldson, MCATD, CTDP, CMP, PMP, PRP

There are very few firsts left in aviation. From the first hot-air balloon in 1783, the first glider in 1848, and the first powered flight in 1903, aviation has progressed at Mach speed, breaking the sound barrier and venturing into space in just over 60 years. So what's left for a young pair of engineers to secure their place in aviation history as pioneers?

Human-powered ornithopter!

Orna-what-now? Okay, let's break this down. An ornithopter is an aircraft that flies by flapping its wings like a bird. Creating a working ornithopter is challenging enough, but creating a human-powered one is almost insurmountable—almost.

In 2008, when a couple of University of Toronto aerospace engineering students showed up at the glider club with plans to build a human-powered ornithopter, I was intrigued but almost certain it was impossible. Then I got to know Todd Reichert and Cameron Robertson, the brains behind this monumental idea, and realized this was going to work. I signed up immediately. While I was unable to help much on the build because of work obligations, I did the majority of the flight training for Todd and was part of the flight crew.

Time to nerd out! There are two main challenges when it comes to a human-powered ornithopter: power-to-weight ratio and how to create thrust from a flapping wing.

There is a good reason why we do not have steam-powered airplanes: There simply is not enough power for the weight of the engine. Hence the reason we did not start flying powered airplanes until the invention of the internal combustion engine, which finally had the power-to-weight ratio that makes sense for flight.

In the case of human-powered flight, your engine is about one-quarter horsepower. This results in the need for a super-light, super-large, and super-efficient airframe to be able to fly with so little power. Enter modern materials. With carbon fiber and Kevlar composites, we are now able to build very large and light airframes. *Snowbird*, the human-powered ornithopter, weighed in at just 42 kilograms with a wingspan of 32 meters (three meters more than the Boeing 737–400).

Now they had to figure out the flapping part. There are two engineering challenges here. First the hinge. You see, a hinge effectively focuses forces and stresses into a single point, the pivot. Any engineer will tell you that when you have focused stresses, you need heavier structures to manage those forces. And so I refer you to the paragraph above.

On the ornithopter project, components were measured in grams. The night before the record flight, holes were drilled in the trailing edge of the tail to save seven grams of weight, illustrating how weight-critical this project was.

The solution to the hinge problem? Get rid of the hinge! This became known as the hingeless moment. The moment that Todd and Cam realized that they needed to get rid of the hinge, they started down a path of using modern materials flexible enough to enable the flapping of the wing.

But flapping up and down is not enough. While that will produce lift, you also have to produce forward thrust. As a bird flaps its wings, they are tilted up on the upstroke and down on the downstroke. This is a complex organic movement that is difficult to replicate mechanically. Again turning to flexible materials, the team added a small extension to the trailing edge of the wing, at its tip, that aerodynamically caused the wing to tilt correctly through the flap cycle.

With the aircraft built, we started flight testing. First flights were straight glides. We would tow the aircraft into the air and release it, and it would glide to a landing. We could fit three of these flights down the length of the runway. When the flapping flights began, it was easy to see the difference: We were going much farther down the runway than before. Because the craft was so light and so large, we had to fly when the air was still, zero wind. Most test flights were conducted either very early in the morning or just before sunset.

In the end they were able to prove sustained flight. We measured the speed and altitude at the beginning of the flight and again at the end. As long as we showed the same or greater altitude and speed, there was energy going into the system.

With this achievement, Cam and Todd turned their sights to another even more elusive aviation first, a human-powered helicopter. In 1980 the American Helicopter Society issued this engineering challenge, one so great that the engineering community has long held it to be impossible—something that is a good exercise in theory but not attainable. Challenge accepted!

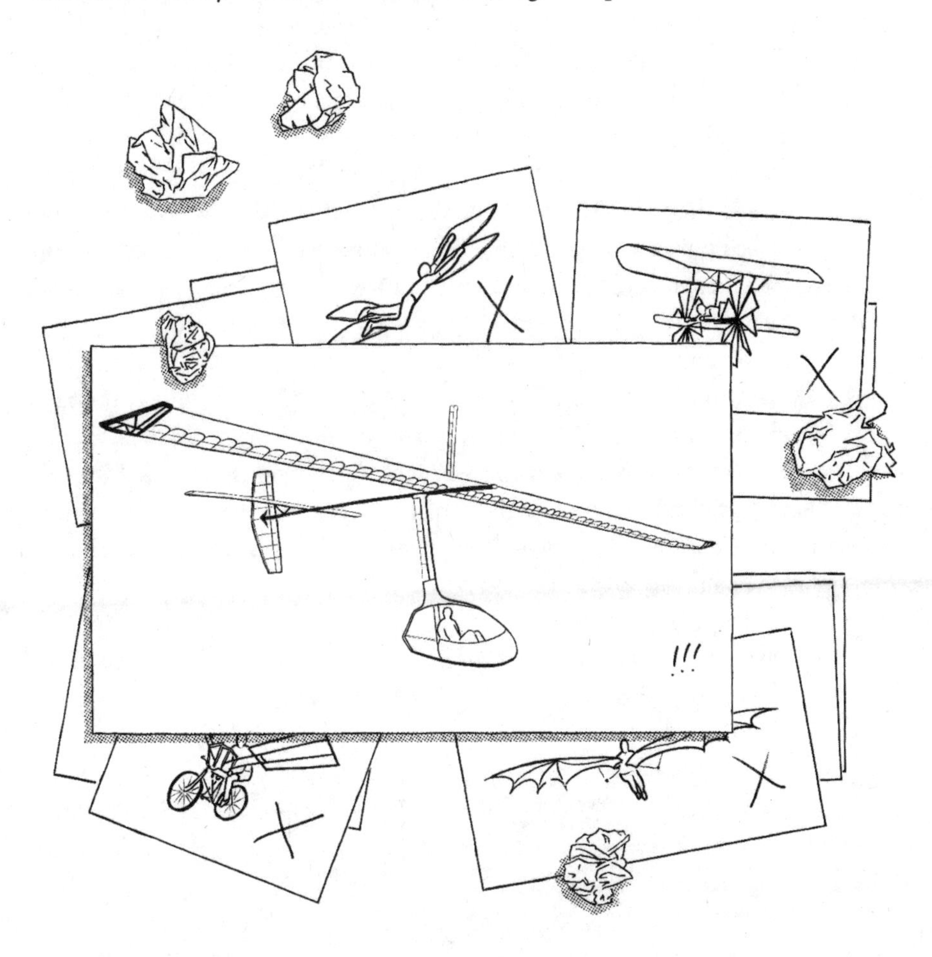

Helicopters are inherently inefficient, not to mention mechanically and aerodynamically complex. Todd and Cam started by running some numbers and found it was theoretically possible. The craft needed would be big, like really big, like bigger than a Boeing 777, while weighing about half as much as your average adult passenger.

So they set about to build a quad-copter where each rotor was 20 meters in diameter, making the entire craft over 50 meters across. The frame was later shortened and the rotors overlapped to address a structural flex issue. At the center was a bicycle that was connected to each of the four rotors. As Todd pedaled, four strings were pulled in, unwinding from the rotor hubs to drive them. At 14 rpm they were not spinning quickly and with a finite amount of string had a very tight window of flight, about two and a half minutes. That was more than enough for Todd!

One of the main principles at play here was to move a very large amount of air slowly. The alternative, of course, is to move a small amount of air very quickly. The team's decision was easily made.

So all you have to do is build a quad copter that is nearly 50 meters across and weighs less than the pilot flying it! With structures that were so thin and light, we had to use special care whenever handling any parts of the craft. Grip a tube of the truss with the same strength you would grip a pen and you would break it. Although it only weighed 55 kilograms, it took four people to pick it up and carry it, one at each rotor.

The design and build would prove to be extraordinary. Through the flight testing there were a couple of crashes; lessons learned, the craft was rebuilt and adjusted until we triumphantly achieved what was previously deemed impossible. I am so grateful to have had these truly once-in-a-lifetime opportunities to work with the team that made not one but two historical firsts in aviation.

David is a professional trainer, facilitator, speaker . . . amateur nerd, and three-time Nerd Nite presenter. He is also a glider pilot (thirty-seven years), flight instructor (twenty years), and Aerovelo team member for ornithopter and helicopter projects.

NERD NITE ST. CLOUD

WHAT DOES GOOGLE SEE?

by Michelle Henderson

It can be tempting to think of Google as a fuzzy warm blanket friend with all our best interests at heart. After all, it was Larry Page and Sergey Brin, founders of Google, who figured out how to create a helpful portal into the internet, almost like people who created a helpful map for the American Wild West. And while Google wasn't the first search engine, it was, and remains to this writing, the most useful.

Google helps people find anything. From the best local dentist to the closest place to buy socks. People love and trust Google. However, Google can also be a mysterious nemesis to most businesses, displaying competitors and negative reviews for popular searches. Hence, many businesses struggle with Google.

So why do—or should—we care what Google sees?

Google's search algorithm is a closely guarded secret. But for the past 25 years (at the time of this writing in early 2023) Google has had one main goal: to present the most relevant response to any user query. As technology has matured, the approach to that goal has shifted and matured. For example, several early search engine optimization (SEO) strategies are irrelevant or harmful today:

- Domain keyword stuffing (BestOrlandoPlumber.com)
- On-page keyword stuffing or keyword blocks (bar, pub, nightclub, club, dance club, meeting spot, et cetera)

- Text colors that match background colors
- Link circles/exchanges
- Paid links
- Links from blog comments
- Syndicated content

Why? Because, over time, Google learned that actual people using the internet don't like those things. They're spammy, not user-friendly, and make a website harder for a human to navigate, read, or use. What people don't like, Google doesn't like.

The challenge is how Google quantifies and weighs what people like in its algorithm. And now that software can learn our individual patterns and behaviors, each person's search engine results page (or Google search experience) is highly individualized to location, device, and previous behavior. In essence, we create and curate our own Google search results through our actions.

Put on the Google Glasses (Not the Google Glass. No One Liked Those.)

Google quantifies and weighs our actions, combines them with billions of actions from others, and then uses that information to predict exactly what we will find most helpful. In that practice, Google has specific tools to use—all available to a software that "reads" text and "thinks" in numbers.

Text on the Page

This is the text that you see as an individual when you're reading a website. The order of the text and how it is tagged in the code (p, H1, H2, and so on) is important. So is the quantity of text in relation to the quantity of code. The quality of the text is also important. Unique, well-written, compelling copy in paragraph form with a modest reading level is going to beat out bullet lists, duplicate copy, dense copy, or "fluffy" copy.

Text in the Code

There's a huge amount of text in a website that users don't see. Within this code, the load order, relevant tags, special markup tags, and metadata are all important signals that communicate with search engines.

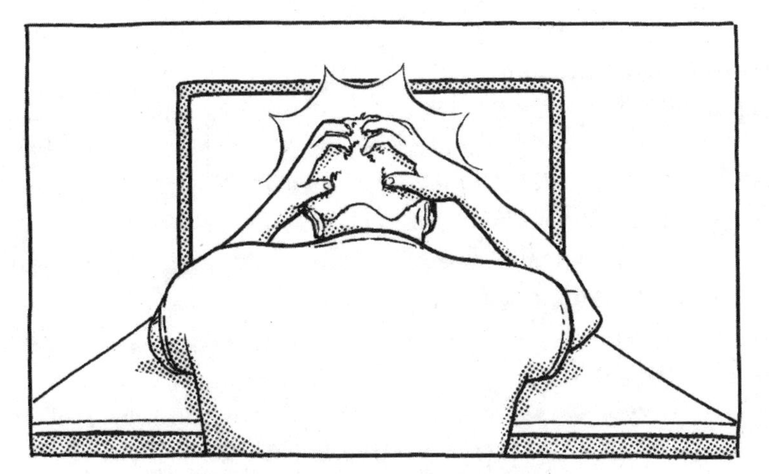

WHY DOES THIS SEARCH
ENGINE HATE MY WEBSITE?

Text Behind Images and Videos

Google is working on quantifying visual information embedded in still images and videos for search algorithms. However, at this time, that information is communicated via text used to describe media files, including the file name. This also happens to be useful information for those who use screen reading software (typically visually impaired or neurodivergent folk), so making these robust and descriptive is doubly important.

Load Speed

With more and more mobile users every minute in every age group and demographic, average load times are hugely important. People simply won't wait three seconds or more for a website to load anymore. Our expectations are immediate. Which means a search engine's expectations are the same.

Other Domains That Link to Yours

These popularity votes are incredibly important. If your content is so helpful to users that another website is willing to link to yours, Google pays attention. There are no shortcuts left here. Paid links, link exchanges, and spam link sites are all actively penalized now. To earn links from other websites to yours, you'll have to get helpful and creative.

URL Structure and Keywords

The words before the .com (or .org or .co or whatever) used to matter. A lot. And then for a while no words in the entire URL string mattered. And now the words and characters *after* the .com matter quite a bit. Why? Because Google recognized that people aren't interested in navigating an entire website that contains the answer to their query. People want to click on the page that holds the answer so we can learn it faster. So when you click a Google organic listing from the search engine results page, you aren't likely to land on the website's homepage. You're likely to land on an interior page.

What Can a Business or Organization Learn?

If your only interaction with Google is as an individual, then most of this information is simply that—mildly interesting information good for party small talk, impressing first dates, and beating your friends at bar room trivia.

If you interact with Google as a website developer or owner, the context of these ideas is important. Because it is so easy to fall into the habit of thinking about your website like a book, read from page 1 to page 40 and printed in stasis, never changing unless some information is outdated. Often, a sentiment like, "Google hates my website for some reason. And the other guy doesn't even have product XYZ and Google loves it," is more about whether Google can "see" and "read" the information than anything else.

Websites that people love are websites that Google loves. When people click a link from Google to a website, then click the back button on the browser and choose something different from the search results page, they're telling Google they don't like that website. When people search on Google, don't choose anything from the list, then search again, they're telling Google that the websites were not what they wanted. When that happens billions and billions of times a day (which it does), Google is paying attention to which websites are showing or not showing at those times.

If your website is one that people leave or skip, the ranks go down. If it's one that people love and return to, the ranks go up. The secret to great SEO is to know how Google quantifies that information today and adjust your plan accordingly. While your website needs to be attractive to people, you also need to know how Google interprets what people find attractive.

Michelle Henderson started BadCat Digital to help small business owners navigate the digital marketing space. With a team that is experienced, creative, and growing, BadCat helps clients discover creative ways to communicate their brands online with transparency and integrity. Her MarioKart *score is the lowest at BadCat, but she perseveres.*

NERD NITE TOKYO

BECOMING A CYBORG THROUGH DISABILITY: Building Prosthetic Limbs

by **Rafaela Libano**

There is something about exceeding the limits of our bodies that humanity finds so enticing. Cyberpunk stories regularly create super-strong humans with infrared vision or mechanical wings, and we wonder if we'll ever reach a day in which that could be part of our reality; we similarly cheer on innumerable superheroes who have mechanically altered their bodies to allow them to fly, punch, and kick infinitely better than they ever could without them. Technically, however, we already kind of have that. Let's talk about prosthetic limbs for a bit, shall we?

Disability has always been taboo. Most of the things around us are created for able-bodied individuals and the expectation is that everyone *wants* to be able-bodied. Why would you want to be broken? Well, as Hugh Herr, professor of media arts and sciences at the MIT Media Lab and a double lower-limb amputee, would say:

> *A human being can never be broken. Technology is broken.*
> *Technology is inadequate.*

With how technological our society is, we cannot keep thinking that it is the human who must adapt to the world around them. If that were the case, we wouldn't have created bridges, buildings, elevators, cars. The same idea applies to a person with a disability. The barrier isn't the disability—it's the space around them or the products that are inadequate to their needs. That's why the World Health Organization's 2011 *World Report on Disability* defines barriers as "factors in a person's environment that, through their absence or presence, limit functioning and *create disability*—for example, inaccessible physical environments, a lack of appropriate assistive technology, and negative attitudes towards disability." To reinforce Professor Herr: Technology is broken. Fortunately, technology evolves. And this is where prostheses come in.

Prostheses started out to serve two purposes: to help a person's movement and to hide limb or facial differences. For example, the oldest prosthesis known to date is an artificial toe, known as the Greville Chester toe. It's estimated to date back to before 600 BC and is believed by researchers at the British Museum to have been used by its owner for regular tasks. And though the Greville Chester toe was functional, up until the nineteenth century, many people sported prosthetic noses to hide the damage done by syphilis before the discovery of penicillin—in other words, for aesthetics. Eventually, prostheses started becoming more functional by mimicking the limbs' natural movements to give the wearer "a more normal life," as prosthetic legs would bend at the knee and prosthetic arms could open and close their hands. They didn't start out being great at mimicking a real limb's functions, but they evolved as time went on to be particularly good at their jobs.

Nowadays, we have very high-tech prostheses that are the envy of many able-bodied individuals. I have heard numerous people jokingly (as far as I know) say they would chop a leg off to wear that real cool prosthesis they saw on the internet. There are robot hands that can thread needles and running blades that can make you run faster than your meat legs ever could. What does the future even hold?

Traditionally, prosthetic limbs are handcrafted since each residual limb is different and needs a bespoke prosthesis. The prosthetist needs to take a cast of the limb, fill it with plaster, and create a test socket out of plastic to make sure it fits the wearer well. After testing, they follow the same steps and continue to finalize the socket with carbon fiber or fiberglass lamination and acrylic resin to make it sturdy. This process is time-consuming because it is crucial that the socket be as comfortable and resistant as possible—it's the part of the prosthetic limb that is in contact with the wearer all day, every day. After a few months or years, the wearer should technically get fitted again, but, since health care is quite a confusing and expensive thing in our world, that doesn't always happen.

Luckily, technology continues advancing and we keep looking into alternative ways to make this process easier. With the use of 3-D scanners and printers, prosthetists can work together with engineers and designers to create more replicable prosthetic limbs faster. This is particularly useful for children, who can outgrow their limbs in a matter of months. As 3-D printers evolve and become more resistant, they could become a viable way to lower costs and time when creating limbs. Meanwhile, science (and other engineers, what *can't* they do?) also keeps advancing from a different front: by developing prostheses that can *feel*.

In 2017, a woman lost her arm and underwent a surgery called targeted sensory reinnervation with the Johns Hopkins Applied Physics Laboratory. The surgery grafted the nerves that previously connected to her hands to a new location on her upper arm. Those nerves were then connected to a sensory cap attached to a prosthetic limb that gave her the ability to feel with the prosthesis. It's been years and this still blows my mind. Being able to control a prosthetic limb with your mind was already wild, but also to have sensory feedback like with your body? Do you know who can do that?

Cyborgs. We keep thinking they are a thing of movies and video games, but they are humans and a part of reality. The future is now!

Rafaela is a product designer who has worked with industrial design and now focuses on UX and UI design for digital products. She is passionate about accessibility and has always made sure to include that in her work, be it in her research on prosthetic limbs and other mobility devices or in her current work on accessible digital products.

NERD NITE AUSTIN

MACHINE LEARNING FOR A FREE AND OPEN INTERNET

by Dr. Brandon Wiley

My work is making tools to help people all around the world access credible and relevant information during increasingly frequent incidents in which the internet is partially or completely blocked as a form of political oppression. Although you may live in a country where this is so rare that you don't even know what I'm talking about, most people in the world experience internet censorship and even complete internet shutdowns on a regular basis.

While many people experience the internet primarily from a wireless device using Wi-Fi or cellular, or perhaps even access the internet via satellite dish, those are just the first links of the connection. The internet then immediately goes straight into a wire and travels around the world from wire to wire. In fact, the weakness of the internet is that it's mostly made of wires.

If you see a telephone pole, these days it's most likely carrying the internet rather than a telephone signal. So how does the internet get across the ocean to other continents? Well, believe it or not, it actually goes through huge cables that carry the signals across the bottom of the ocean. There are only a few such cables, which is a problem because they frequently get destroyed by the anchors of ships catching on them, thereby disrupting the internet for a whole country at once. The cable has to enter the country somewhere, so this means that whoever controls that location also controls access to the internet and therefore can decide what does and does not go through that cable. Frequently, adversaries will block things starting with articles that are critical of that adversary, then whole websites, followed by sites on which one can freely post content such as chat apps and social media, and then tools that let you circumvent the blocking such as proxies and VPNs. The blocking gets increasingly restrictive, and sometimes regimes just unplug the internet entirely.

My dissertation work was on what I call network protocol shape-shifting, which is a way to circumvent this blocking and restore the ability to access information. Except when the adversary blocks all network traffic, there are always two categories of traffic: allowed and blocked. The adversary must have some way of differentiating network traffic into these two categories. Accomplishing this at the scale of an entire country's internet requires that it be done in an automated way, using machines that have rules to classify the traffic. The machines are programmed by humans, but the humans cannot directly observe the vast flow of information. Only the machines are fast enough to do so, and they just execute the rules they are programmed to use. Therefore, in order to get your traffic through, you just need to figure out the rules and follow them, landing your traffic a classification into the allowed category.

This is where machine learning comes in.

Machine learning is a general term for algorithms that create mathematical models based on observations. In my definition, what I mean by *algorithms* is computer programs, what I mean by *mathematical model* is a data structure, and what I mean by an *observation* is some bit of data. This may sound like a wild oversimplification, but I want to demystify this idea for you.

At its core, machine learning is just a way that we write computer programs to process data into a data structure. Once we have this data structure, we can use it for two things: classification and regression. Classification takes new data and puts it into learned categories. Regression takes a category and generates new data that would fit into that category. Machine learning can't do anything beyond the realm of what computer programs in general can do. The code that machine learning algorithms run is written by people, the data is gathered and input by people, and the results are interpreted by people. It's fundamentally just a form of computer programming. Machine learning is still very cool, though, as it's a particular way of writing computer programs that can help solve problems that are hard to solve just by thinking about them really hard yourself.

In my dissertation work, I used machine learning algorithms to look at observations of traffic from the allowed and blocked categories and generate a mathematical model representing the probable rules used. Once I have the rules, it's relatively simple to come up with a way to transform existing network

traffic so that it follows all of the rules necessary to get an allowed classification. In fact, machine learning can help by performing regression for us.

Regression is when an algorithm generates predictions of potential observations for a category. Thus, giving the machine learning algorithm the allowed category, it can generate example network traffic that would likely be classified as allowed by the blocking rules that it has learned through observation. In a sense, we are asking what "normal" traffic looks like, and fortunately for us, "normal" must be given a rigorous technical definition that the machine can understand. You can think about it like this: The machine can only observe a limited set of characteristics of network traffic, those it has been preprogrammed to know how to analyze. For each of these characteristics, there is a subset of values that are considered acceptable and a non-overlapping subset that are considered unacceptable. All we need to do is produce acceptable values for all observed characteristics and we're in the clear! It's easy to overthink the problem, dreaming up all sorts of complex characteristics that traffic could have. However, at the scale of a whole country's internet traffic, the calculations must be kept extremely simple. Therefore, the transformations that we have to make to the traffic are pretty simple as well. Even more fortunately, the machine learning does all the work for us!

Someday I hope that we will have machine learning systems that are automatically trained to transform our traffic transparently so that blocking becomes impractical. We just have to build it!

Dr. Brandon Wiley has worked for over twenty years in the internet freedom space on projects such as Freenet, BitTorrent, and Tor. The doctor is a Google Ideas Research Fellow and a contributing author to Peer-to-Peer: Harnessing the Power of a Disruptive Technology. *The doctor is a founder and the president of Operator Foundation, a 501(c)3 nonprofit organization that specializes in bespoke technology development for supporting global human rights.*

NERD NITE SAN FRANCISCO

HOW TO WIN FRIENDS AND INFLUENCE BACTERIA

by **Sarah Richardson, PhD**

Humans have domesticated cats. And dogs. And fish. And birds. But could, or should, bacteria be next?

Bacteria run this planet. Their influence is easy to overlook because we refer to their actions vaguely. For instance, we say, "*I am gassy today*," not "I fed my gut bacteria fiber and they made me fart," or "*Cow burps heat up the planet*," not "The cow's microbes made methane that the cow vented." Or "*My orange rotted*," not "I waited too long to eat my orange and microbes ate it first." But we could not exist without them.

Bacteria let us breathe: One bacterial genus called *Prochlorococcus* lives in seawater and is responsible for at least half of our oxygen. We must keep the oceans healthy to keep them producing breathable air.

Bacteria feed us: Most of the atmosphere is nitrogen, which is great because we need nitrogen in many molecules in our bodies. But atmospheric nitrogen is inert, so what we breathe in we breathe right back out unabsorbed. Only bacteria can convert inert nitrogen to bioreactive nitrogen, which they feed to plants (which we eat for nitrogen). Without bacteria, the only sources of edible nitrogen would be lightning strikes and rock erosion.

Bacteria feed us more: Sometimes when food changes color and stinks, it's tastier. Fermentation is the process of feeding microbes and harvesting their . . . product. Alcohol might be the oldest example: Feed yeast sugar, get ethanol.

Bacterial fermentation gives us coffee, chocolate, vinegar, cheese, salami, olives, MSG, hot sauce—and more!

Bacteria heal us: Antibiotics, dewormers, pesticides, and antifungals are molecular tricks borrowed from bacteria to fight pests and pathogens. The best way to understand your enemy is to learn from *their* enemy.

Bacteria clean up after us: The carbon we ingest but don't digest gets excreted; the carbon we don't ingest gets chucked into compost piles; and when we die, we are sometimes buried. Bacteria break down all abandoned carbon to building blocks for return to the planetary cycle. Without bacteria, we would be hip-deep in dead stuff.

Bacteria let us ruin the planet: Petroleum and coal sprang from the microbial carbon cycle. All petroleum on the planet is dead life at least partially processed by microbes. All coal on the planet is the result of dead life that microbes couldn't process fast enough. The type and quality of petroleum is determined by what died and who ate which bits. Call this a toss-up—oil is useful but bacteria didn't force us to build our economy on it.

Bacteria can do even more: Forget the whole humans-are-the-most-accomplished-form-of-life-on-the-planet thing. The list of bacterial skills above is nowhere near exhaustive, and when it comes to chemistry, they have us licked. Humans use unsustainable petroleum to make almost every chemical we use in daily life, but bacteria make the same molecules from the crud in compost heaps. There are bacteria that produce hydrazine—rocket fuel—from wastewater. Some species make magnetic material more perfectly formed than any human has ever managed. Some make the basic building blocks used to assemble materials we think of as solely sourced from petroleum.

Could we prevail upon their generosity and achieve a more sustainable modern world?

Domestication

There is a proven tool for collaboration with our planetary cohabitants. Domestication is the process of making a contract with another life-form. You say, "I will protect you from disease, predators, and the elements. No need for you to grow horns or make pigments for camouflage. I will change how I organize my society to include you. All you must do is specialize in something you're already good at. Do it more, do it better."

And with a "moo," the deal is struck. You build barns and fence-in pastures, stockpile hay, and strike an accessory deal with herding dogs. Over time the cows get more docile; their horns get smaller and their coats get whiter. They make more milk and breed more often. A successful bargain is of mutual benefit. You and the cow are both more assured to pass on health and success to your offspring. The cow's genome changes under artificial pressure; your genome changes with ready access to cow resources.

Domestication is the greatest engineering success in human history. We have dogs, corn, almonds, goldfish, turkeys, strawberries, horses, broccoli, chickens, honeybees, grapefruit, and all the rest because bioengineers recognized useful skills, learned new languages, and brought partners from the wild to the farm. Success required humility and a unique respect for each organism.

Although we sometimes recognize their skills, we have been neither humble nor respectful toward bacteria.

Domesticating Bacteria

What's the holdup? You may have heard of it: *Synthetic biology* is a buzzword for technology deployed to exert control over life. It is currently the dominant paradigm in microbial bioengineering. Practitioners think of bacteria as simple machines that can be "programmed with DNA." They believe the skills of one microbe can be genetically transferred to another, and that this is necessary because some species are too inconvenient to work with. It's disrespectful. You would never genetically engineer goats to prey on mice just because cats are aloof. Manually rewiring an herbivore to eat meat is staggeringly more difficult and miserable for everyone than genetically engineering friendlier cats; it only seems easier if you don't care what goats or cats think. Uncompromising behavior is guaranteed to be unsuccessful. There is no controlling life; even the tiniest organisms have opinions.

Bacteria are neither simple machines nor ancient, inscrutable life-forms. They're as young as we are, and many are as ready to strike a bargain as any other organism we've met. If we approach them the same way we approached the wolf and the wildcat, we could both benefit, and on a much-abbreviated timescale!

To make friends with bacteria, we must respect them enough to strike an honest domestication deal. When we promise to deemphasize petroleum and commit to a bioeconomy where they feel protected and useful, they'll help us save the planet.

Sarah (the germ wrangler) is the CEO of MicroByre, a company dedicated to the domestication of bacteria. She holds a PhD in Molecular Biology and has a long history of advocating for bacterial rights.

NERD NIGHT PITTSBURGH

WHY NUCLEAR FUSION WOULD BE AWESOME—IF WE GET IT TO WORK

by **Dr. Matt Moynihan**

I have the power!" We know those as the immortal words of He-Man, but if he were living today trying to create fusion, he'd probably instead have to say, "I *almost* have the power!"—which, admittedly, doesn't quite have the same effect.

So why such an endeavor? Why He-Man? And why fusion?

Because reliable fusion would be the greatest power source ever harnessed by humanity. Simply, nuclear fusion is the process by which two light atomic nuclei combine to form a single heavier one while releasing massive amounts of energy, and if replicated on Earth (it's what powers the sun), it could provide virtually limitless clean, safe, and affordable energy to meet the world's energy demand.

Fusion is the Holy Grail of energy. It would be a seemingly endless supply of zero-carbon power. It cannot melt down. One fusion reactor would be capable of replacing several of our current nuclear plants. And it's abundant; fusion fuel can be extracted from seawater. In fact, a single bathtub full of fusion fuel could meet all of a person's energy needs for 30 years. Wow!

But man's goal of re-creating and harnessing fusion is not new.

The first time we got bulk reactions was in 1957 during the Zero Energy Thermonuclear Assembly (ZETA) experiment in the United Kingdom, and as we flash-forward to today, over 200 hobbyists have achieved fusion in their homes and garages. However, despite each of them successfully creating energy, none of them have been net positive—the community's Kitty Hawk moment will be the first time we get more power out than we put in. But we're getting close.

As of 2022, that record is 59 megawatts of fusion power made from a 100-megawatt driver. Humanity is almost there. And as fusion has gotten closer, it has moved from a public sector endeavor to the world of start-ups.

More than 30 start-ups around the world have popped up over the last 25 years and they have collectively raised more than $5.3 billion in private funding, with investors like Bill Gates and Jeff Bezos aggressively backing their favorite approaches. As with any new technology sector, most will fail, but the odds that one of these teams reaches net power in the next 10 years are dramatically better than they were only a few years ago. And not a moment too soon, either, because as climate change wreaks havoc on our cities, farms, and ecosystems, we need fusion power to stave off the worst impacts of climate change.

Which brings us to magnets.

The key technology that is making all this possible is the development of superconducting magnets. Superconductors can make magnetic fields continuously with no resistance, and in some fusion devices, the rate of fusion energy scales at the strength of these fields to the fourth power. Hence, when superconductors are added in, a whole new generation of fusion machines emerge. The reactor goes from a pulsed device to a continuously running power plant; it morphs into a machine that can make a great deal more energy.

Moreover, more than half the devices being pursued by start-ups use magnetic fields in some way, so, this is an industry-wide trend. Until the late 2010s, the best fusion devices could muster was one to three Tesla fields* using water-cooled copper magnets. But today there are companies planning for all superconducting machines that reach 10 to 20 Tesla. And still stronger fields may be possible, as the research team at the DC Field MagLab in 2022 proved a 45 Tesla magnet suitable for fusion devices.

Finally, fusion rockets are the coolest technology that I can think of. Yes, rockets. Of fusion.

In a fusion rocket, a thruster puts a fusing plasma inside the exhaust cone of the rocket engine. Materials leaving the fusing plasma stream away at some percentage of the speed of light. When they strike the cone, they are deflected

* For those of you not working on fusion in your spare time, a Tesla (T) is the International System unit of field intensity for magnetic fields. One Tesla (1 T) is defined as the field intensity generating one Newton (N) of force per ampere (A) of current per meter of conductor. Got it? Matt and Chris had to look this up, too.

out into space and push the rocket forward. The resulting reaction is so powerful that we could get to Mars in a matter of weeks using this kind of thruster. In fact, there are a handful of US fusion rocket companies that have built small prototypes of these thrusters using NASA funding. Fusion rockets are a subset of the larger fusion industry, and they will also get a huge performance boost when superconductors are added. Anyone interested in learning more should reach out to the Fusion Industry Association, Rocket Committee, for more information (and that's one to grow on).

You yourself may now be at net negative energy after reading this, so simply understand that fusion power is great for fighting climate change and fusion power plants cannot melt down. However, though fusion works well as a clean baseload energy source, it cannot supplant the role of other renewable power sources like wind and solar, which we still desperately need to deploy at scale. Ultimately, fusion power is an important part of this renewable-power balanced breakfast.

Dr. Matt Moynihan has been involved in nuclear fusion since 2006. He holds a PhD from the University of Rochester, was a nuclear engineer for the US Navy, and authored a popular science book on fusion for Nature-Springer. He was the host of The Fusion Podcast *and* The Nuclear Fusion Shark Tank. *In 2018, he started New Light Fusion Consulting LLC, which helps investors understand this $5 billion industry.*

Math Is Fun

From ages six through 17 I collected baseball and football cards. And for the most part, that was all I spent my allowance and minimum-wage-job income on. And I still have all 100,000 cards today sitting in boxes and binders in a bedroom closet. While I only flip through them a couple times each year, I fondly hold on to them because they make me recall one important thing they taught me—math. Look at the back of any card of any sport and you'll find stats. Some of them are very countable, whole-number stats such as home runs, RBIs, or touchdowns that a player had in a given year, but there are also plenty of stats that are decimals, such as earned run average, batting average, yards per catch, and shooting percentage. And this is how I gained the ability to calculate fractions and percentages in my head—quickly.

For example, a pitcher's ERA is: 9 × earned runs / innings pitched. Therefore, if a pitcher gives up one earned run in nine innings, their ERA is a neat 1.00 (9 × 1 / 9). But what if a pitcher gives up three earned runs in two innings? That's a 13.50 ERA. Or what if a hitter went 132-for-500 in a season. That means they hit a respectable (for these days) .264. As a kid who spent thousands of hours staring at the back of baseball cards and price guides, fractions and percentages became second nature and helped me tremendously my entire life. Who knew that packs of thin cardboard and stale gum could be so practical? Though I stupidly avoided a career that involves complex math, I can still multiply and divide large numbers quickly and correctly in my head. While not useful every day, it makes for good party tricks or tip calculations at dinner. So that's something.

Whether you're a sports-stats nerd or not, the coming pages are a blast. You'll get to know infinity and how our cities can improve by simple math. You'll learn where codes and cryptography came from, how classical and pop music would make a mathematician blush, or simply how you talk about your besties behind their backs. Math is everywhere, didn't you hear?

—Matt

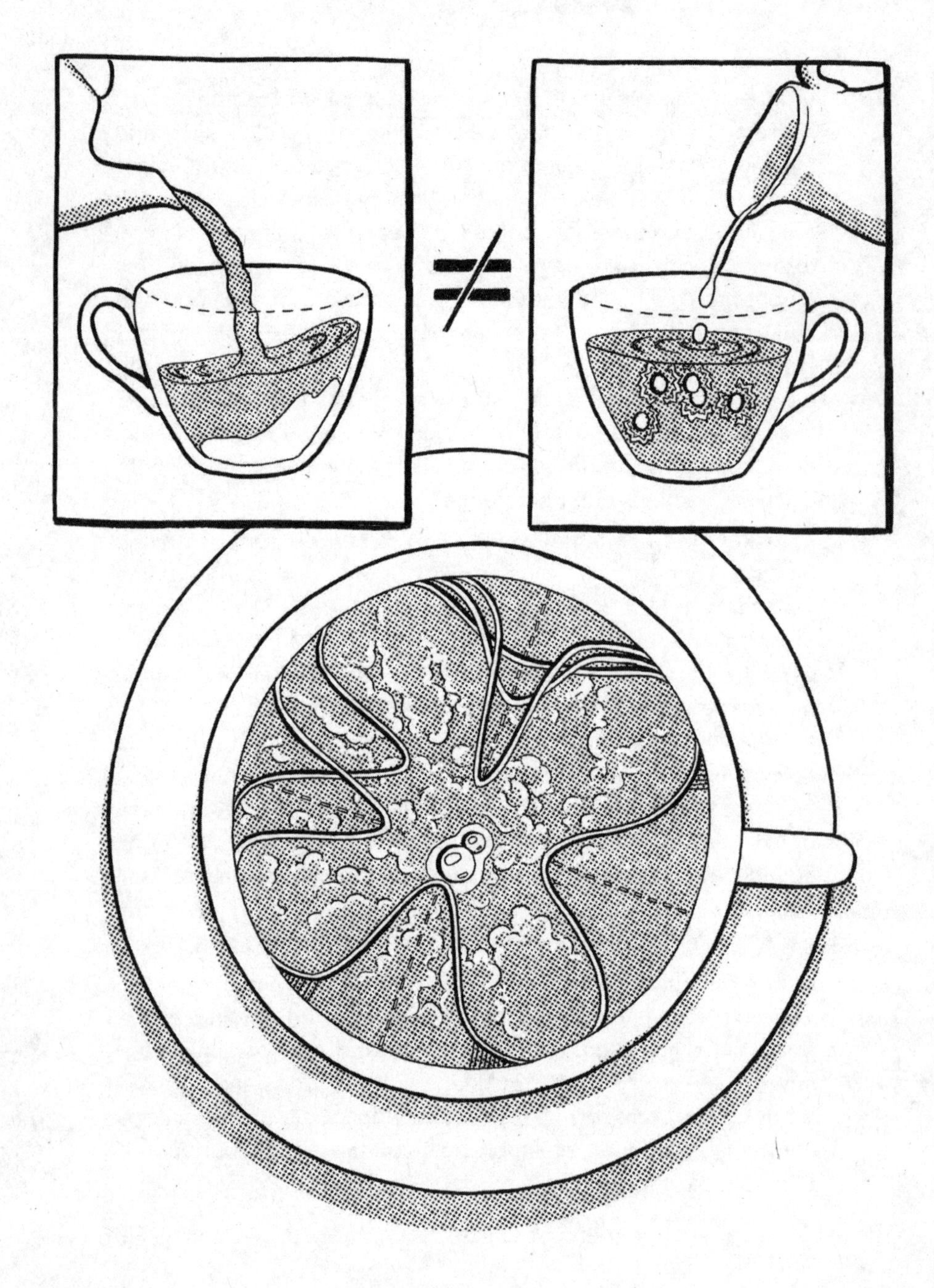

NERD NITE DC

A TEA TEST TEMPEST

by **Sam Kean**

When Ronald Fisher offered his colleague tea that day, he was just being polite. It was the 1920s, in England. What could be chummier than a cup of tea? Fisher had no idea he was about to kick up a big fight—much less revolutionize modern science.

Fisher was a short, slight mathematician with rounded glasses. He worked at an agricultural research station north of London. He'd been hired to help scientists there design better experiments, but he wasn't making much headway. The station's tea breaks were a nice distraction.

One afternoon, Fisher offered to make a cup for biologist Muriel Bristol. She said yes, so Fisher poured some milk into a cup and added tea. That's when the trouble started. Bristol refused the cup. She declared that she never drank tea unless the milk went in second.

Now, this debate about tea—whether to add milk to the cup first, or tea first—has been a bone of contention in England for centuries. Each side has its partisans, who get boiling mad if someone makes a cup the "wrong" way.

But as a man of science, Fisher thought the debate was nonsense. It's simple thermodynamics. Mixing A with B was the same as mixing B with A. How could adding milk or tea first make any difference? He told Bristol to stop being silly.

But Bristol insisted that the order mattered. She even claimed she could taste the difference between a milk-first cup and a tea-first cup.

Fisher scoffed, claiming that was impossible. So Fisher and Bristol made a bet. He proposed making eight cups of tea, four milk-first and four tea-first. He'd present them to Bristol in random order and let her guess which were which.

Bristol agreed, so Fisher disappeared to make the tea. Upon returning to the room, he found an audience of other scientists gathered. They wanted to see whether Bristol knew her stuff.

Fisher presented the first cup. Bristol sipped and smacked her lips. Then she made her judgment. Milk-first.

Fisher handed her a second cup. Tea-first, she said.

This happened six more times. Tea-first, milk-first, milk-first again. By the eighth cup, Fisher was goggle-eyed. Bristol got every single one correct.

How? Well, it turns out that adding tea to milk is not the same as adding milk to tea, for good old chemistry reasons. Milk contains fats and proteins that are hydrophobic, or water-hating. When those water-hating molecules encounter hot water, they get scalded and change shape, curling up into balls.

That's important because a change of shape results in a change of flavor. Specifically, the proteins acquire a slight burnt-caramel taste. This is especially likely to happen when you pour milk into hot tea since the first drops of milk get isolated and are therefore more likely to get scalded. In contrast, if you do the opposite and pour tea into milk, the milk drops never get isolated. Little scalding occurs, and there's little production of off flavors like the burnt-caramel taste.

As for whether milk-first or tea-first tastes better, that depends on your palate. But Bristol's intuition was correct. The chemistry of milk ensures that milk-first cups and tea-first cups taste distinct.

Overall, Bristol's triumph was humiliating for Fisher. He was proven wrong in the most public way. But the important part of the incident is what happened next.

After his embarrassment, Fisher wondered whether Bristol had gotten lucky. Maybe she'd just guessed correctly all eight times. Admittedly, this was petulant of him. But he worked out the math for this possibility. And he realized that the odds of her guessing correctly eight times in a row were 1-in-70. Thus, she could probably genuinely taste the difference.

But even then, Fisher couldn't stop thinking about the experiment. What if she'd gotten just one cup wrong? He reran the numbers and found that the odds of her guessing "only" seven of eight correctly dropped from 1-in-70 to around 1-in-4.

In other words, if she'd missed just one cup, she probably could still taste the difference, but Fisher would have been much less confident. And what intrigued Fisher is that he could quantify exactly how much less confident he'd be. He could put a number on it.

Furthermore, that lack of confidence told Fisher something: that the bet's sample size had been too small. He should have made her try more cups.

So he began running more numbers and found that 12 cups of tea, with six each way, would have been a better experiment. Each individual cup would have carried less weight, so one data point wouldn't skew things as much.

Before long, other variations on the experiment occurred to him, like using random numbers of tea-first or milk-first cups. He explored these possibilities over the next few months.

Now, this might all sound like a waste of time. It's just a cup of tea. But the more Fisher thought about it, the more profound this work seemed.

You see, in the 1920s, there was no standard way to design scientific experiments or analyze data. In fact, his agricultural research station hired Fisher to work on those very problems, and again, he hadn't made much progress. But he realized that the tea test pointed the way. However frivolous it seemed, the simplicity of the test clarified his thinking and allowed him to isolate the key points of good experimental design and good statistical analysis. He could then apply what he'd learned in this simple case to messier, real-world examples—like, say, isolating the effects of fertilizer on crop production. The same general principles would apply to experiments in physics, chemistry, and biology, too.

Fisher published the fruit of his research in two highly influential books, which introduced ideas that scientists still use today, like the null hypothesis, statistical significance, and the danger of small sample sizes. And the first example in the first book—to set the tone for everything that followed—was Muriel Bristol's tea test. Over the next century, Fisher's extrapolation from that bet revolutionized modern science, influencing how every lab in the world works today. Not bad for a guy who couldn't make a proper cup of tea.

Sam Kean is a New York Times *bestselling author of six books, including* The Disappearing Spoon *and* The Icepick Surgeon. *He and his work have been featured on NPR's* Radiolab, Science Friday, *and* Fresh Air, *and his podcast debuted at number one on the iTunes science charts. http://samkean.com*

NERD NITE BOULDER

THE MATHEMATICS OF GOSSIP

by **Izabel Aguiar**

Mathematics is marvelous at modeling many different aspects of our world; from transportation systems and tumor growth to brewing the perfect cup of coffee, mathematical modeling provides us an in-depth and beautiful way of understanding processes that happen all around us. Beyond applications in physics, chemistry, and biology, mathematical modeling has long helped us understand psychological and sociological processes, too.

One common mathematical modeling method is the SIR model, which is often used to model the spread of a disease, as this model computes the theoretical number of people infected with a contagious illness in a closed population over time. In fact this method was used quite regularly in 2020 to predict the spread of COVID-19. We began using SIR models to study epidemics and the spread of infectious diseases in 1960, but in 1964 researchers William Goffman and Vaun A. Newill began using these same modeling tools to track not diseases, but ideas—they called it the "transmission of ideas." So how does this apply to gossip?

There is now a wealth of research on studying the spread of information in a community: More modern, complex models use tools from network science, sociology, and statistics, even incorporating data from social media. The original, most basic models use *differential equations* to represent the interactions between different people in a population, as well as other additional parameters, to account for the infectiousness and rate of recovery for a given disease.

Without even writing down these differential equations, we can still understand the models through a few examples.

To explore how a piece of gossip spreads, let's assume there are three parts of a given community: the susceptible (S), the infected (I), and the recovered (R)—thus, SIR. The susceptible are people who haven't heard the gossip, the infected are people who have heard and believe the gossip, and the recovered are people who have heard but no longer believe the gossip. Different pieces of gossip have different parameters describing their qualities about how believable and infectious they are. Depending on the gossip, people in a community can switch between being susceptible, infected, and recovered at different rates.

To understand the effect of parameter values, mathematicians often do *sensitivity analysis*: checking how different combinations of parameters affect the overall behavior of a system. When thinking about gossip, we can do a more fun, *anthropomorphized sensitivity analysis*. Let's think about gossip spread by three different gossipers representing a different combination of parameter values: Regina George (Rachel McAdams's character in *Mean Girls*), Dr. Neverheardofher, and a Conwoman.

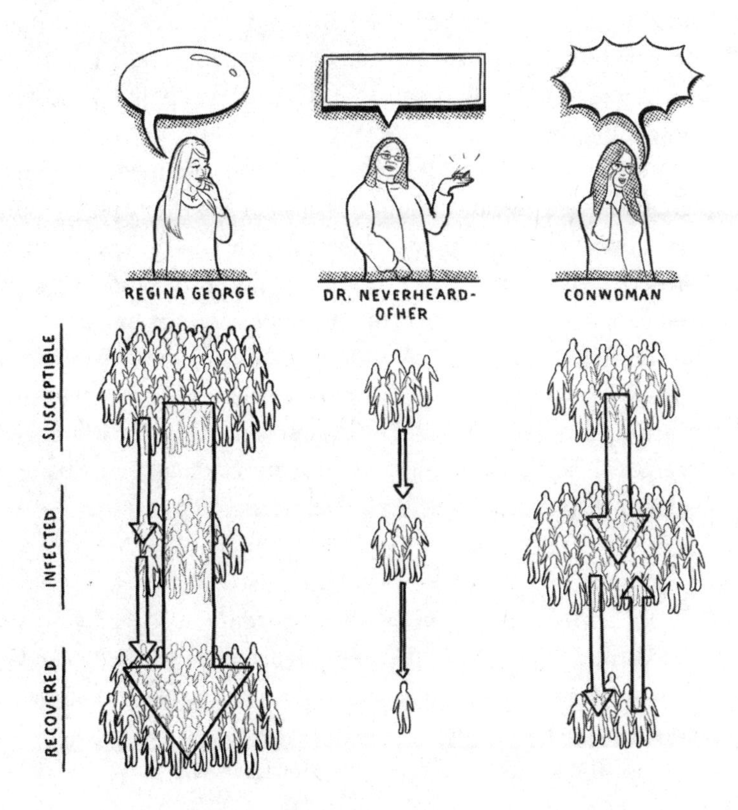

Regina George is popular, has a lot of influence in her school, gossips about things her peers care about, but is well known for being untrustworthy. She's currently spreading gossip about who-brought-whom to a party last weekend (Like, can you believe that Sandy and Pat came together? OMG!). In our SIR model, even though Regina's popularity and relevance mean that this gossip spreads at a high rate, most people immediately reject it (recover) without ever having been infected, and those who do believe it recover quickly.

Then there's Dr. Neverheardofher, an esoteric but well-respected entomologist, and she's decided to spread some gossip about one of the crickets she studies ("That damn cricket OU812 was chirping all night!"). Whereas this gossip doesn't immediately reach far or spread very quickly (I mean, who's actually interested in gossiping about crickets?), not many people are immune to the gossip because she's *the* expert. Moreover, because she's so well respected, the rate at which people recover is very low.

The Conwoman is a person with moderate influence and whose gossip is topical. Moreover, the Conwoman is so charismatic that even people who have recovered from her gossip can become reinfected by interacting with people who still believe it. Perhaps the gossip she's spreading is ideological or political ("Did you hear that the mayor is merely a puppet of Big Nerd Nite?"): It takes longer to be infected, but it is so strong that those who believe it want to continue spreading it to others.

Comparing the spread of gossip from these three gossipers, we find that whereas Regina George's gossip very quickly reaches the whole population, after a few days only a very small proportion of people still believe it's true. Dr. Neverheardofher's gossip takes about seven and a half times longer to reach the whole population, but over 95 percent of people continue to believe it even after a month has passed. Finally, even though the Conwoman's gossip takes longer than Regina George's to spread, the population becomes increasingly infected: The gossip remains persistent even after some people have recovered.

These three gossipers show us that the *source and content* of gossip are critical in understanding its persistence and influence in a community.

Izabel Aguiar is a PhD student at Stanford University, where she works with Johan Ugander to develop and use tools to understand complex networks. She's excited about teaching, baking interesting cakes, and telling her family and friends how much she loves them.

NERD NITE NYC

FROM BACH TO TOOL: The Secret Math Behind Music Theory

by Alexander Brewer

What is the golden ratio?

Glad you asked. With a jargony topic at the intersection of math and aesthetics, it's important to begin with a solid and specific definition. The golden ratio, which is also referred to as the golden mean, divine proportion, or golden section, exists between two quantities if their ratio is equal to . . . the golden ratio.

Does that clear things up? No?

These abstruse mathematical formulas should help! In a case where $a > b > 0$,

$$\frac{a+b}{a} = \frac{a}{b} = \varphi$$

$$\varphi = \frac{1+\sqrt{5}}{2} = \frac{a}{b} = 1.618033988749\ldots$$

To put it simply: The golden ratio is found between two numbers if the ratio of the smaller number to the larger is equal to the ratio of the larger number to the sum of the two numbers. It's often approximated by the Fibonacci sequence (1,1,2,3,5,8 . . .), where each number is the sum of the previous two (3 + 5 = 8). As you go higher and higher in the Fibonacci sequence, the ratio between the last two numbers approaches this irrational number (987/610 = 1.61 etc. etc.).

Even if you're not a numbers person, you are probably already familiar with

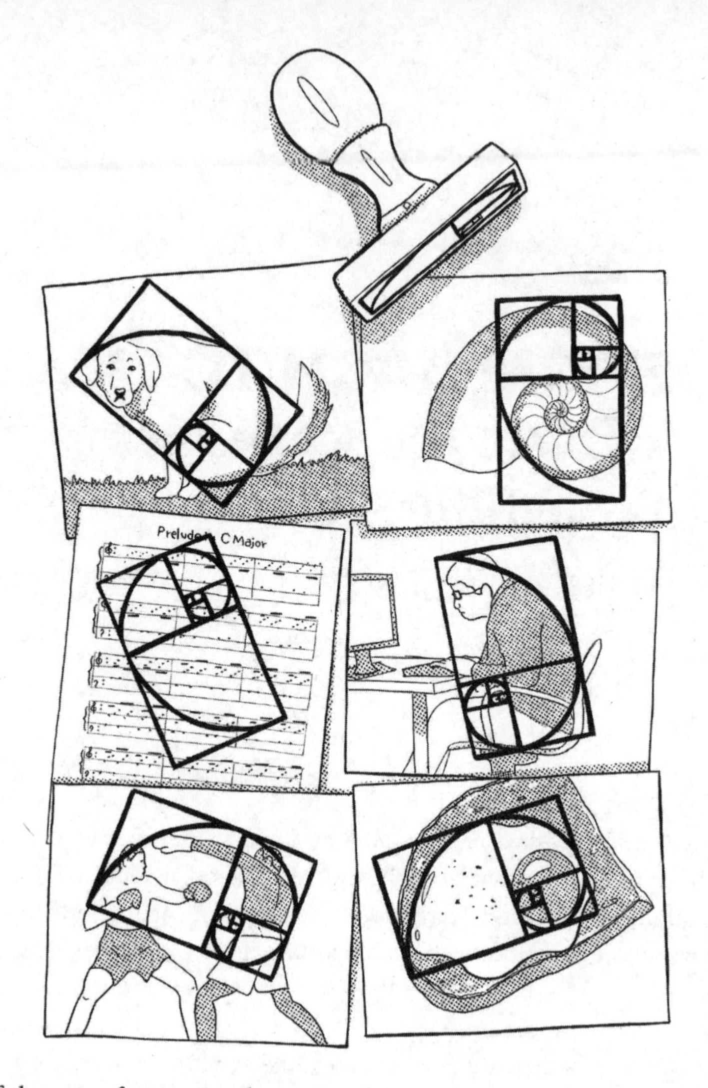

one of the most famous implications of the golden ratio: the "Golden Spiral." (*Editors' note: Neither Chris nor Matt is familiar with this, so don't feel bad if you aren't, either.*) It turns out that if you have a series of squares each related by the golden ratio, when you connect a circle drawn inside of each square to the larger square next to it, you get this really visually pleasing spiral shape.

And we know that it's both visually pleasing and useful because it shows up in art and in architecture—and, in fact, it even shows up in plant biology. In a paper studying the way that leaves arrange themselves around the stem (a property known as phyllotaxis), mathematical biologists (yes, that's a thing) Bergeron and Reutenauer were able to show that these patterns follow the golden ratio. This ratio also shows up in our own bodies. In a research paper from the *International Journal of Cardiology*, researchers Yetkin, Yalta, and Yetkin found that the average diastolic to systolic blood pressure is very close to the golden ratio!

But if this ratio is so important in other disciplines like art and biology, does it also present itself in music?

First, we'll look at some of the physical properties of sound. Most musical sounds are not just the frequency that we hear as the "pitch" of a note, but also many other frequencies. This series of frequencies is known as the harmonic series, and the note that we hear is referred to as the fundamental. There are an infinite number of "overtones," or harmonics, sounding with that note above the fundamental. Going in order, the 2nd-order partial is an octave higher than the fundamental, the 3rd is an octave plus a fifth, the 4th is a fourth higher, 5th is a third higher, and so on. The space between two or more of these notes is known as an interval.

You can take any two numbers and create a ratio from them, but let's look specifically at harmonics corresponding to the Fibonacci sequence. A lot of these ratios do correspond to sonorous intervals—in fact, the notes of the pentatonic (1:1-> Root, 8:5-> Third, 3:2-> Fifth, and 2:1-> Octave) are all Fibonacci numbers.

We also see the Fibonacci numbers in the scale degrees themselves. There are eight notes in the octave scale and 13 notes in the chromatic scale. If we put these scale degrees side by side on a piano, we see that the numbers that line up from those two scales corresponding to the Fibonacci sequence are C, D, E, G, and C, which make up the major pentatonic scale.

In addition to the physical makeup of music, we see various composers throughout the ages drawing inspiration from the golden ratio. One composer whose music's hidden mathematical structures have long fascinated researchers is J. S. Bach. The Fibonacci numbers show up in the bar counts of entire works, such as *The Art of the Fugue*, as noted in *The Mathematical Architecture of Bach's* The Art of Fugue. We also see it in the structure of individual pieces. For example, in the Prelude in C Major from Bach's *Well-Tempered Clavier, Book I*, the Fibonacci numbers appear in the actual note groupings. The piece is rhythmically built around a five-note ascending pattern followed by a three-note ascending pattern. There are also several interesting major 7th chord moments in the piece, with each one occurring in a golden ratio section. Modern researchers such as Michael Linton point out this was an odd chord for that time period, and it's no coincidence that Bach places these chords there.

And what do you think we find at the actual golden section of the entire piece? Bach's signature, B♭, A, C, B♮.

The golden ratio was not exclusive to Bach, however. In the rhythms that Béla Bartok wrote in his third movement of *Music for Strings, Percussion and*

Celesta, we see reference with the Fibonacci numbers in the note groupings again. But the ratio is not exclusive to a particular time period—we also see it in modern music.

Progressive metal band Dream Theater's "Octavarium" heavily references the Fibonacci numbers in the artwork and time signatures. The 1980s band Genesis's "Firth of Fifth" uses the Fibonacci numbers in the number of measures of guitar solo in this piece. And we see this in Tool's "Lateralus" as well. The golden ratio is a heavy inspiration in this song. The first thing we might notice about this song is the Fibonacci numbers in the verse's syllabic content.

Verse:

Black (1)
And (1)
White are (2)
All I see (3)
In my in-fan-cy (5)
Red and yel-low then came to be (8)
Reach-ing out to me (5)
Lets me see (3)

In the chorus, too, we see a set of three time signatures (9/8, 8/8, and 7/8) that, when combined, are equal to a number in the Fibonacci sequence (namely, 987). A similar phenomenon happens in the bridge, but instead of consecutive time signatures, we actually get two time signatures at the same time (6/8 and 10/8), a technique known as polymeter. Interestingly enough, when you divide these two numbers, you get a very close approximation to the golden ratio!

$$987 \div 610 = 1.618\ldots$$

How many of you reading this are thinking to yourselves, "This guy sounds like he's full of shit"? I mean, this is kind of . . . a stretch, right? It feels like we're "over-thinking, over-analyzing." It kind of seems like this heady math stuff is "withering our intuition." Like we're just "reaching for the random" or "whatever will bewilder" us. In fact, those are literally the lyrics of the song.

Tool knows its fanbase well. They know that their listeners will spend time diving deep into the potential mathematical meaning of the song and ignore the lyrics that are right in front of their faces. And even though I presented research to you, I ended up doing the same thing. I spent a good deal of time preparing this contribution looking past the immediate flaws in some of the research in order to find what I knew I wanted to present on. Looking back, I realized that some of the articles I referenced are poorly written. The ideas that many of these articles propose are narrow-minded, Western approaches to using a number in music. We're taking for granted that an octave is divided into eight parts, or that there's even an octave to tune to!

For example—let's go back to the harmonic series. There's really no evidence that the harmonic series has anything to do with the golden ratio or Fibonacci numbers—they are just coincidences taken out of context. Furthermore, our modern Western organization of notes largely comes from an idea known as equal temperament, which relates two octaves by a ratio of 2:1 (not φ), splitting the spaces in 12 equal parts. This doesn't have anything to do with either the golden ratio *or* the harmonic series.

Authors such as Casey Mongoven propose an entirely different tuning system, one that uses the golden ratio as the ratio for the octave, instead of the simple 2:1 ratio we find in the harmonic series. I was able to find some music that uses similar tunings proposed in a paper of his, and they sound, well, pretty bad. This doesn't help our "aesthetically pleasing" argument about the golden ratio at all.

In his work "Stria," John Chowning modulates one frequency by another frequency, specifically selecting the golden ratio as the ratio between the two frequencies. While this is interesting music, I certainly wouldn't argue that it constitutes the most "beautiful" music due to it being based on the golden ratio. To our Western ears, the intervals actually sound horribly mis-spaced.

And while I'm not going to argue that composers like Bach or Bartok didn't use the ratio or that their music isn't beautiful (they did and it is), what really intrigued me as I was researching this was how desperate we were—*I was*—to find this golden ratio in everything.

This isn't anything new—theorists have been discussing the ratio for centuries. In response to music theorists' attempts to find and use the golden ratio everywhere, Gustav Theodor Fechner writes of one theorist in particular:

> *Because he is of the opinion that the same principle of pleasantness must be valid in both the field of vision and sound—and does not want to hear of anything at all that deprives golden ratio of power—he finds himself obligated, in clear contradiction to his general judgment, to declare . . . the golden ratio, to be the musically most pleasant proportion.*

So I'm left with this question: Why? Why do we do this? The golden ratio, at its core, is about relationships between entities. Specifically, what came before, is related to what is, is related to what is to come. I think we find meaning in this number, not because it makes things more beautiful, but because it hinges on the idea of self-reference. And we love self-reference. We see ourselves in our ancestors—my father was a pastor, as was my grandfather, and I, too, love describing my beliefs on stage. We see ourselves in our lineage—our children's faces, their likes and dislikes. In *Gödel, Escher, Bach: An Eternal Golden Braid*, Douglas Hofstadter talks about how he suspects that our ability to self-refer is what gives rise to consciousness. We can see in recursive algorithms that we love to teach computers to do it, too.

Music is also about relationships between entities. In medieval academics, the subjects of math, geometry, music, and astronomy were paired together because they dealt with values, with music being about the relationships and ratios between numbers in time. And we see plenty of evidence of this. We love listening to Bach when the music references itself, such as in the subjects of fugues. We love it when music itself references work by other musicians, such as in remixing and sampling. We especially love it when musicians reference each other or even inspire each other—like when Eminem reminds everyone about Dr. Dre.

The word *pareidolia* is defined as the tendency for perception to impose a meaningful interpretation on a stimulus so that one sees an object, pattern, or meaning where there is none. I think the reason that we find the golden ratio in music, even when there is none, is because we see self-reference in music, we see ourselves in the self-reference, and we see meaning in ourselves. This is nothing to be ashamed of. It's one of the most natural things for a human to do.

Now, do I think that the golden ratio makes music "more beautiful"? Maybe, maybe not. But do I think that finding this number that represents self-reference in the music that we already love, even if it might not be there, makes us more beautiful? Absolutely!

Xander Brewer holds a BA in Computer Science and Music from Boston University and is currently a grad student at NYU Steinhardt studying music technology. He works remotely as a software security something-or-other (which is usually what people hear when I talk about what I do anyways so we might as well cut out the middleman, right?). He may or may not be related to NYC-based spoken word artist and musician The Ringer.

NERD NITE PORTLAND

GETTING TO KNOW INFINITY

by **Dr. Zajj Daugherty**

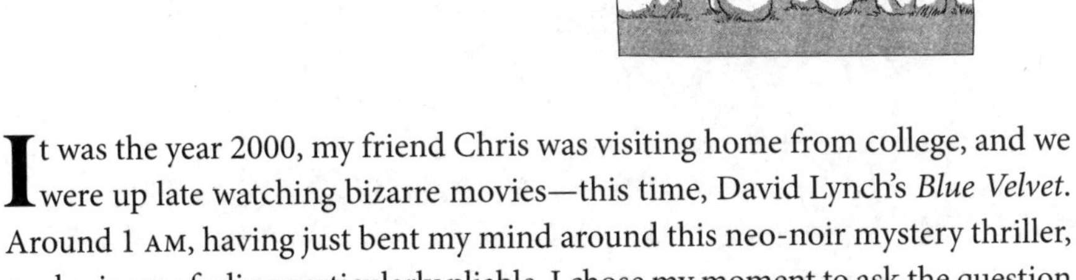

It was the year 2000, my friend Chris was visiting home from college, and we were up late watching bizarre movies—this time, David Lynch's *Blue Velvet*. Around 1 AM, having just bent my mind around this neo-noir mystery thriller, my brain was feeling particularly pliable. I chose my moment to ask the question that had been bothering me.

Chris was a brilliant young man studying mathematics at a top STEM college. I was, myself, a 17-year-old girl at the height of loathing math—seemingly the painful practice of memorizing formulas and performing rote calculations better left to computers. How could you dedicate your life to this nonsense?

So I asked: "Okay, so you're studying *math*. What's *that* about??"

"Well," he replied, "what do you know about the different sizes of infinity?"

The story he proceeded to tell convinced me. Not just of the ill-defined nature of infinity, but also that math is an intriguing, complex, and creative endeavor. I was hooked. It wasn't until almost 15 years later, after having gotten a few degrees in math myself, that I heard the story of the mathematician behind this mathematics—Georg Cantor—who was called a "charlatan" and a "corrupter of youth" for his efforts. Good old Cantor, corrupting youth into becoming mathematicians 80 years after his death!

So what **is** *infinity?* The first thing you need to know in preparation for this story is that infinity is a *theoretical* concept.

Informally, when we think of infinity, what actually comes to mind tends to be really, really big or small physical things. The infinite expanses of the universe. The infinitesimal particles in our surroundings. But these things are all *measurable*. And if you can measure something, then it is finite. When we try to imagine infinity, we carry biases that are weighed down by physical intuition. So if our job here is to compare infinite things, we are necessarily entering a thought experiment that cannot be physically explored. This exercise bends our physical intuition to breaking points. But that's okay! We're taking the freedom of being mathematicians for the moment and running with it!

Learning to count

We begin modestly: You're out for a walk and you come across some groundhogs. How many groundhogs do you see? If your answer is three, you come to it by matching up the numbers 1, 2, 3 to your furry new friends. Every groundhog gets a number; every number gets a groundhog. If you had more groundhogs, you'd use more numbers. Lots more groundhogs—lots more numbers. And though (we hope) there will only ever be finitely many groundhogs around to count, we can be confident that we'll be ready to count them because the counting numbers themselves are infinite!

So now let's pretend that we do have an infinite number of groundhogs, one for every counting number. This is the same thing as lining this overabundance of rodents up with the numbers 1, 2, 3, 4 . . . but never reaching a stopping point. This is what we call countably infinite.

Countably Infinite ***a***

Capable of being matched one-to-one with the positive integers (aka the counting numbers).

This is one of our most familiar kinds of infinite—a sequence of things just marching along ad nauseam. And "infinity + 1 = infinity" still works here. If three more latecomers show up to this groundhogapalooza, we can just ask everyone to move down a bit to make room: original furball #1 moves down to slot 4, original furball #2 moves to slot 5, and so on (the counting numbers never run out of space because they're infinite); the new folks slot into places 1, 2, and 3.

What about infinity × 2? For example, is the set of *all* integers countably infinite? Sure! We need a new strategy, but there's still plenty of room. Instead of asking everyone to move down by the same amount, we can instead shuffle the newcomers in by asking the old groundhogs to move to twice their original spot. (*Now* it's a party!)

Now that we're warmed up, we're ready for a more involved example: the rational numbers.

Rational Numbers ***n***

Those numbers that can be expressed as fractions of integers (e.g., 2/5,-17/3, 0/1).

The trickiness is that the rational numbers are quite crowded: In between any two different rational numbers, you can always find another—between 0 and 1/2 is 1/4; between 1/4 and 1/2 is 3/8; and so on. There's no such thing as a "next" rational number when ordering by size. So we need a new technique!

Start with a grid of (positive) fractions, where the columns index numerators and the rows index denominators. Next, discard fractions that aren't in their "lowest form" to remove duplicates (for instance, 2/4 is the same as 1/2). Starting from the upper left corner, zigzagging your way down, counting as you go. In this way, we find that the rational numbers are also no bigger than the integers they contain!

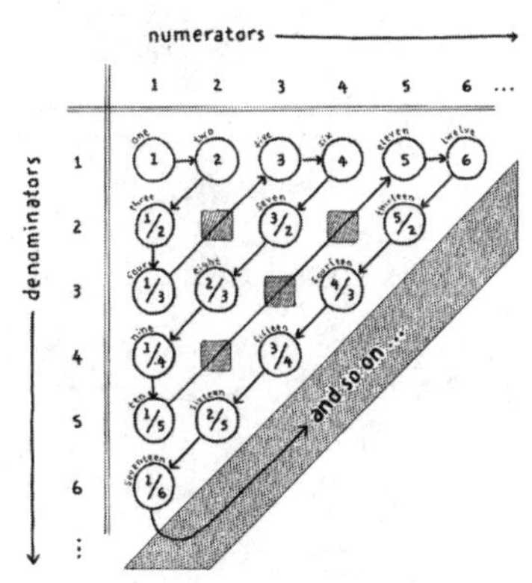

So is there anything that's *not* countable? Yes! For example, the real numbers:

Real Numbers ***n***

Numbers obtained by appending an infinite decimal expansion to an integer.

In order to establish that something is *not* countable, it is not enough to see that all our old techniques won't work. (We've been pretty creative about counting things so far, after all!) Instead, we must show that no matter how much coffee we've had, no matter what genius moments we have in the shower, *any* list of real numbers will be incomplete! Given any list of real numbers, we must devise a plan for finding a number that's not on that list.

The plan: Cantor's diagonal argument

Start with any list of real numbers, each number written with its full decimal expansion:

$$1 = 1.0000000\ldots$$

Moving down that list, circle the first decimal place in the first number, the second decimal place in the second number, and so on. To build a number that is different from every number in that list, work one decimal place at a time:

- If the first number has a first decimal place equal to 0, give your number a first decimal place of 1; if not, give it 0.
- If the second number has a second decimal place equal to 0, give your number a second decimal place of 1; if not, give it 0.

And so on.

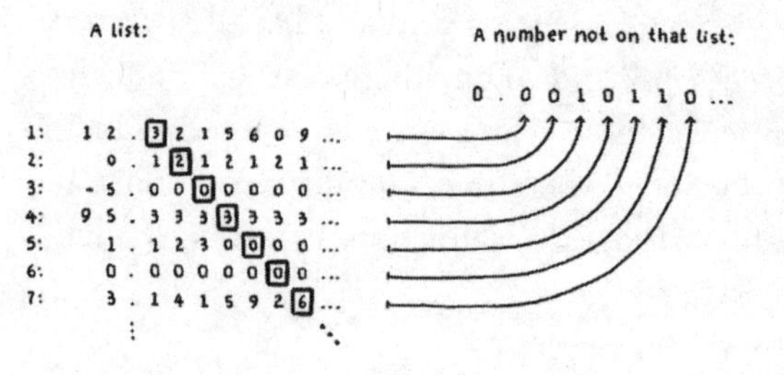

Since real numbers don't care about whether decimals reach an end or follow any kind of pattern, you have successfully built a real number that is different in at least one place from every item on the list!

This was the huge breakthrough—the thing that got Cantor called a charlatan and a corrupter of youth. *Infinity* is not a well-defined term.

Is there more?

We began with the finite: a modest party of three groundhogs. Then we got the first taste of the infinite: the countable. *Then* we found out that there was something bigger: the set of all real numbers. Two different infinities.

But is there more? Is there something out there that's *bigger* than that? Oh my, yes!

If you take the sets of real numbers (like {1}, {1,-3}, {1,-3, √2}, and so on), *those* make up a larger set than the real numbers themselves! And not to be pathological, but it goes on from there: By considering all sets of items in the previous set, you can keep stepping up in size indefinitely. (Think about people in a city, cities in a state, states in a country, and so on.)

There are at least countably infinite many distinct sizes of infinity!

Is there even more?

If that's not enough for you, I humbly invite you to read more about the continuum hypothesis and the controversial evolution of axiomatic set theory in the twentieth century.

Zajj Daugherty is an associate professor at Reed College in Portland, Oregon. She's a theoretical mathematician, studying abstract algebra and combinatorics. Having been raised by artists, though, she has devoted her career not just to pushing on the boundaries of mathematical understanding, but to making math more understandable and approachable for everyone.

NERD NITE ORLANDO

MATH FOR A BETTER CITY

by Eliza Harris Juliano

I grew up in two dramatically different places: suburban South Carolina and hyper-urban Manhattan. When I was eight, we left our big yellow house with a two-car garage and spacious yard for the quintessential big city. Powered only by our feet, we could reach restaurants of every flavor, movies, school, church, ballet classes, and even the opera. I quickly knew the subway map well enough to serve as navigator for my mom when we left our neighborhood. In exchange for our spacious backyard, I had Central Park. By age 12, I regularly walked with schoolmates to the diner half a block away. I could imagine the next chapter of my life in the city being one of quickly expanding independence limited only by my allowance and my mother's consent.

Then we moved back to our suburban homestead. Suddenly that big yard seemed a little smaller. Without a driver's license, everything was too far away. The closest restaurant, a Subway, would have been a 20-minute walk down a 45-mile-per-hour road with a ditch and no sidewalk. The movie theater, a 35-minute walk, might as well have been on Mars. I survived on rides from my parents and older friends until I finally got my driver's license in the spring of my junior year of high school.

Later, in college, I learned that business and government leaders in the 1950s dreamed of a future with gleaming highways connecting neat suburban homes with downtown office towers. With the backing of the federal government this dream quickly became a reality driven by the profits of World War II, the invention of federal mortgage insurance, new engineering standards to enable speedy car travel, and zoning codes that promoted single-family subdivisions. People of color were largely excluded from buying into the dream, first by racist contracts and policies and later by skyrocketing prices, fueling segregation and wealth gaps that persist to this day.

For a while suburbia seemed pretty good for those who got in. But everyone's desire to live on a quiet street meant the streets didn't connect. The shiny new roads got congested, leading residents to eschew new neighbors. Neighborhoods were segregated by increasingly small differences in lot size, a proxy for income. The dream left out anyone too young, old, or poor to drive. Later, researchers linked "suburban sprawl" to myriad modern ailments including pollution, traffic deaths, obesity, wildlife decline, tight city budgets, and teen depression.

Most important, I learned that how we build our cities is a choice. We can choose to again build places that are close, connected, and comfortable:

> **Close:** People are close to places they want to go. This means buildings are closer together. Businesses, schools, and civic institutions are near homes or below apartments.
>
> **Connected:** Sidewalks, bikeways, streets, and transit systems connect to one another, enabling people to choose a variety of methods to travel directly and efficiently.
>
> **Comfortable:** Close your eyes and remember being on a sidewalk that felt comfortable and where you could also get a cup of coffee or stop into a store. Buildings were probably close to the sidewalk with interesting things to see and you felt protected from moving vehicles.

It's not just big cities that accomplish these goals. My subdivision was an annex to the historic town of Summerville, where homes stand on narrow streets arranged neatly in a grid that flows uninterrupted into Main Street with shops, restaurants, and town hall. Many popular vacation destinations have this character, from Paris and Miami Beach to Nantucket.

If we know how to build better cities, why don't we? One obstacle is dysfunctional math:

Street space: Streets are usually evaluated by how many cars they move instead of how many *people* they move. People-based math accounts for walking, biking, carpooling, and transit. In congested areas, where destinations are close and space is finite, a street can move many more people than cars.

The environment: If new housing will be built on ten acres of land, someone who cares about the environment might say, it's better to have only ten houses on that land instead of fifty because ten houses will leave more green space than a denser development. But if fifty families need homes, where will the other forty houses go? Will they impact another five or ten or forty acres of land with the need for more miles of roads, pipes, and driveways, more time spent commuting, and more fuel wasted?

Your city's pocketbook: To make their budgets work every year, cities need revenue. When a city council approves a brand-new suburban supercenter to be built, they know they can expect a big fat property tax check every year. What they don't always realize is that a collection of smaller stores, apartments, and homes on smaller lots can produce many little checks that add up to more revenue overall. As Charles Marohn says in "The Cost of Auto Orientation" and Robert Steuteville discusses in "Walmart Versus the City," we must look at how much tax revenue properties produce per acre rather than per lot.

Your pocketbook: Prospective homebuyers may hear the adage "drive till you qualify." We tend to focus on the cost of housing, ignoring that cheaper housing often means higher transportation costs. Even if a family calculates that walking for errands and taking transit to work would more than offset higher housing costs, they may not be able to qualify for the higher mortgage. The result is families in remote suburban homes they can't afford to leave.

The math is clear, but change is hard. We've been building suburbia on warp speed for 70 years and it's not going away. What we can do now is ensure the next decades of investment focus on refilling historic neighborhoods, strengthening existing towns, and retrofitting suburban places with trail connections and walkable village centers. If you look around, it's already starting. Is your city changing for the better? Consider writing to your city council member in support of a new bike path or new neighbors. Put on your walking shoes, dust off your bike, and discover your better city.

Eliza Harris Juliano is an urban planner living in Orlando, Florida. Eliza helps communities build and retrofit places to be more sustainable, healthier, and prosperous.

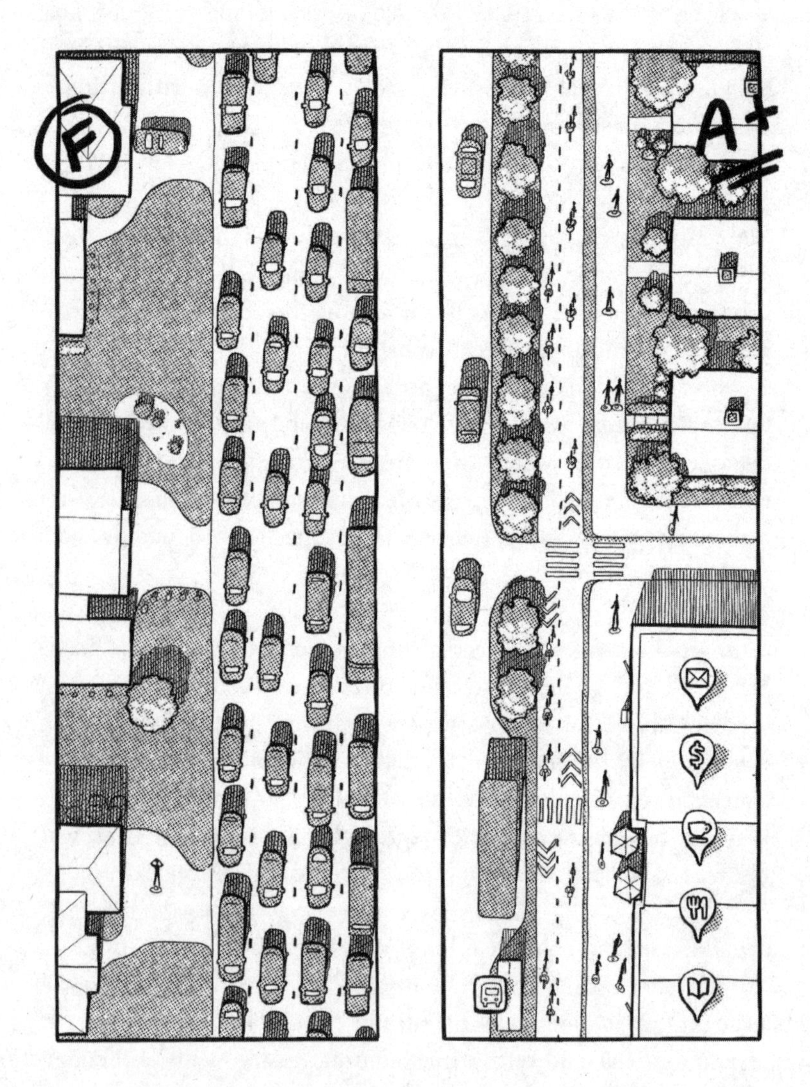

NERD NITE SYRACUSE

A LITTLE "BIT" OF CRYPTOGRAPHY

by mArshaLl sWaTt

People love secrets. They love possessing them, sharing them, and becoming privy to them. People also value their privacy. We know the human need for privacy and secrecy is ancient and crosses cultures because we have evidence dating at least as far back as 500 BC. The evidence exists in the form of primitive substitution *ciphers*, where a phrase is scrambled by substituting a given letter with another letter in the alphabet. This may remind you of the ancient language of Pig Latin. Ipherscay areyay unfay!

This ancient method is still appealing, if impractical nowadays, because it is satisfying to anyone who cracks it. Substitution ciphers represent the earliest form of *cryptography*, encoding a message to keep its contents private. We couldn't survive in the modern world without cryptography, as it most notably enables e-commerce and data encryption (for example, the little lock symbol thingy in your web browser).

Let's go over a little terminology. *Cryptology* is an umbrella term that refers to the study of ciphers and methods to defeat them. A *cipher* is any method or algorithm for encrypting a message. *Cryptography* specifically focuses on ways to create ciphers, while its counterpart *cryptanalysis* focuses on the means and methods to decipher or "crack" an encoded message.

In 1466, a thousand years after the first recorded cipher, Renaissance Italian architect Leon Battista Alberti made the first major improvement by introducing the polyalphabetic cipher, or ring cipher. He devised a slight modification to the substitution cipher by periodically changing the substitution mapping of letters as a message is encoded. This method made cyphertext harder to break than previous approaches.

Figure 1: Alberti's ring cipher

Then, around 1518, Johannes Trithemius conceived of a method to hide an encoded message within a larger plain message—in his case, a seemingly innocuous Latin prayer. Instead of a one-to-one letter substitution, a single letter was encoded by substituting it with a specific Latin phrase (for example, substituting *Rex* for *N*, *Dominus* for *E*, *Fabricator* for *R*, and *Optifex* for *D*). This "Ave Maria" dictionary cipher gave birth to steganography. Prayer has never been the same since.

The next major advance was the use of a secret *encryption key*, a further modification to the substitution cipher method, devised by an Italian named Giovan Bettista Bellaso in 1553. In this method, each letter of the key was numerically encoded following the order of the letters in the alphabet (A= 0 . . . Z = 25).*

```
Key = P  A  S  S  W  O  R  D
    = 15 0  18 18 22 14 17 3
```

* This example was drawn from https://www.cs.virginia.edu/~evans/dragoncrypto/day1.html.

To encrypt a message, you match the key to the message, starting again from the beginning once you reach the end of the key.

```
Message:  A  T  T  A  C  K  A  T  D  A  W  N

Numbers:  0  19 19 0  2  10 0  19 3  0  22 13

Key:      P  A  S  S  W  O  R  D  P  A  S  S
          15 0  18 18 22 14 17 3  15 0  18 18
```

This message is then encrypted by summing numbers associated with the message and the key. When the sum is greater than 26, you count starting back at zero, such that 19 + 18 (37, is encoded as 11). This is known as modular arithmetic on a 26-number scale (mod 26). Using the example above, summing the 0 (from the A of the Message) and the 15 (from the P of the Key), you get 15, making the first letter of the CipherText "P" (below), etc.

```
            15 19 11 18 24 24 17 22 18 0  14 5
CipherText: P  T  L  S  Y  Y  R  W  S  A  O  F
```

The short keys typically used in Bellaso's cipher make it vulnerable to a *frequency attack*, the substitution ciphers' unrelenting kryptonite. In a frequency attack, the known frequencies at which different letters are used are used to decode the message.

But the next improvement would change cryptography permanently.

The first unbreakable cipher, known as a *onetime pad*, was invented in 1888 by Stanford trustee Frank Miller. Miller improved on Bellaso's invention by making two small but critical improvements: (1) using a randomly generated encryption key and (2) making the encryption key at least as long as the message itself. Unbreakable may be a slight exaggeration . . . more specifically the ciphertext is *informationally secure* because there is no way to infer anything about the original message from the ciphertext. Onetime pads were used in World War I.

I'll end our survey of cryptography here because after this, it gets a little heavy for bedtime reading. Cryptography is an elegant mathematical game, played across time by two adversaries, endlessly devising ways to encipher and decipher the many secrets we dream up and share with those deserving of our trust. The next level of this game will likely be the development of ciphers that are safe from attack by quantum computers. If you'd like some further pleasure reading about cryptography, I highly recommend Neal Stephenson's *Cryptonomicon*, which weaves (indeed embeds) detailed descriptions of several cryptographic methods into a fascinating fictional story.

Marshall has worked in the blockchain industry for several years. As a software engineer and manager, he has helped design, build, and manage software platforms for start-ups and large companies. He currently lives with his wife in his hometown of Syracuse, New York.

Careers

Let's face it, for almost every single one of us reading this book, your high school guidance counselor, if you were fortunate enough to have had one, had a pretty limited view of the careers that are out there. In school you probably heard about the standard jobs: postal worker, police officer, lawyer, librarian, nurse, engineer, sanitation engineer, musician, music teacher, some other type of teacher; the list goes on. In this section of the book, we want to introduce you to some career paths you might not have considered: (1) chemistry graduate student turned science educator; (2) mortuary scientist (both historical and modern approaches); and, of course, (3) wildlife detective. We also want to point out that some careers you always thought you wanted, like perhaps being a surgeon (of human patients) or a surgeon (of animal patients), might not be quite what you envisioned. I'll be honest, I thought Dr. Lee M. Bishop was going to write about urine (based on a Nerd Nite presentation for which he's well known), but we are happy to add his inspirational contribution about being a researcher and a friend to this section.

—Chris

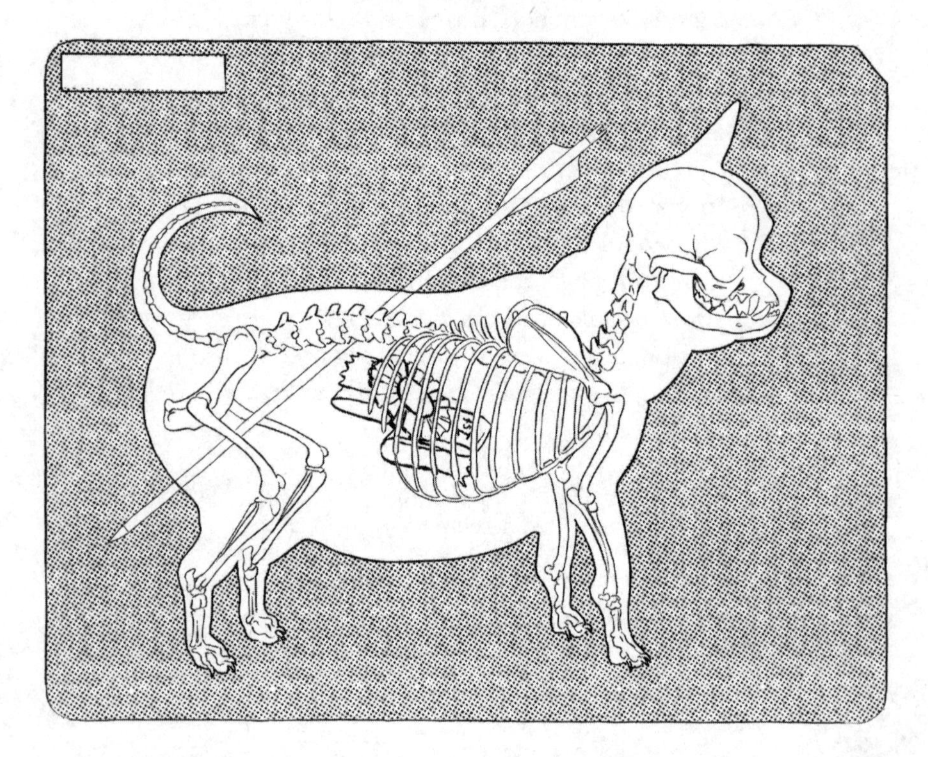

NERD NITE GREENVILLE, NC

VETERINARY CONFIDENTIAL

by Jessica Girard, DVM MS

Growing up, I loved collecting fossils. How I got from there to being a veterinarian is a bit of a twisted tale. I studied plague in prairie dogs, I served coffee, I searched libraries for obscure texts about mosses, I sequenced ant DNA, and I served as a researcher for the *Annals of Improbable Research.* During this last stint, I befriended a veterinarian, spent some time working at a vet clinic, and was hooked. Despite being drunkenly warned at a party by a bitter veterinarian *not* to follow this path, I went to vet school, got braces, earned my diploma, stethoscope, and lab coat; got my braces taken off; and set off to live this irresistible dream. And while the job has been certainly rewarding, little did I know how many darker, stickier, and smellier truths I'd uncover only once I was actually on the job. If only I could borrow Doc Brown's time machine to sit down with my younger self and tell her what I know now. But luckily, I can at least tell you.

If you're wondering if becoming a veterinarian is a good career path, let me say, maybe not, and may I suggest something like accounting instead? You know, a nice steady job where you don't find yourself pulling half-digested used tampons out of a dog's intestines or being sprayed in the face by anal glands.

Here is a short list of things I have made dogs vomit:

A best-in-show ribbon the dog had won earlier. And ate because it tasted good or in silent protest of her show dog lifestyle?[1]

A diamond earring.

An entire pack of bacon and, another time, six (six!) whole pork chops.

Socks.

Many varieties of Halloween, Christmas, and Valentine's Day chocolate candies (chocolate smells good even in vomit—disturbing, but true).

Large piles of raisins and grapes.[2]

True to form, it's actually very hard to make cats throw up. Luckily, they eat fewer stupid things than dogs (but for some reason, hair ties are some cats' kryptonite). Just ask Matt's cat, Lu.

Things that I have seen/assisted with removing in dogs that should not have been there:

Very large pairs of women's underwear. Note: Thongs probably have a better chance of passing through a dog's intestines than granny panties do. Thus, if you're a dog owner, consider buying sexier undergarments, at least for their sake more than yours.

Socks.

The squeakers from toys.

An arrow (surely the person who shot this dog with an arrow is actively pursuing their career as a serial killer now).

1 The owner said, "That little bitch!" when referring to the dog's crime, which was an accurate description since the dog was small and female.

2 If you did not know, grapes and raisins can be weirdly toxic to dogs and nobody knows why.

Shotgun pellets, bb's, bullets.[3]

Here is a quick list of things to do to stay on the good side of your veterinarian:

Please don't tell us we are in it for the money. The twenty-year-old car we have in the parking lot and vast amount of school loans we took out so we could get bitten by your toothless Chihuahua speak otherwise.

In a similar vein, please don't make that joke about how we should name this exam room / part of the hospital after you because you "have spent so much money there."

Do not bully us online or in person. Talk to us instead. Did you know veterinarians have a higher suicide rate than dentists? We know you love your animals; we are here to help as much as possible. But we are also not gods. We can't diagnose your animal without diagnostics, which cost money. But if you have no money, you can tell us that and we will do our best to help with what you have. And yes, sometimes we make mistakes, just like human doctors. Because we are human.

So in summary, just be nice to us. And don't shoot your neighbor's dog with arrows.

Jessica Girard is a veterinarian who enjoys neutering cats and making dogs throw up things they shouldn't have eaten in the first place. She also benefits hugely from nepotism since she's married to Chris Balakrishnan.

3 In fact, our own dog that we adopted when she was a year old has a hidden pellet in her hip that we never knew about until she needed X-rays. We postulate she was stealing chickens or some other ne'er-do-well behavior because she is a consummate ne'er-do-well.

NERD NITE DC

CHINDOGU: The Japanese Art of Unuseless Inventions

by **Josh Manning**

The chopstick fan. A baby mop onesie. Solar-powered flashlights. A toilet paper tissue hat.

Chindogu is the Japanese art of eccentric invention. Often causing more problems than they solve, Chindogu ultimately serve no real purpose. Neither useful nor useless, they are therefore "unuseless" inventions, similar to how "undead" means not dead but also not alive.

Coined by Kenji Kawakami in the 1980s, *Chindogu* is a sort of portmanteau that roughly translates to something like "weird or curious tool or device." Chindogu merges the bizarre absurdity of a mad scientist with the genius and innovation of an aspiring armchair inventor.

While seemingly chaotic and unstructured, Chindogu must adhere to "The Ten Tenets of Chindogu" as defined by the International Chindogu Society:

1. Not Really: A Chindogu cannot be for real use. If it is actually helpful then it is not a Chindogu.

2. Exist-essential: A Chindogu must be created. It can't just exist on paper. If it doesn't exist, then the world will never really know that it is neither useful nor useless.

3. Anarchic: Chindogu is free to be what it needs to be and is broken free from the chains of usefulness.

4. Universally Unuseless: Chindogu are tools for everyday life. Everyone must recognize the uselessness.

5. Not for Sale: Chindogu are not for sale, even as a joke.

6. Stop Trying to Be Funny: Humor must not be the sole reason for creating a Chindogu.

7. Propaganda . . . Not: Chindogu are innocent and are made with the best intention.

8. Keep It Clean: Chindogu are never taboo. They must not represent cheap sexual innuendo, vulgar humor, or sick jokes.

9. Don't Get Greedy: Chindogu are offerings to the rest of the world and cannot be patented, copyrighted, or otherwise privately owned.

10. Chindogu for All: Chindogu are without prejudice. All should have a free and equal chance to enjoy each and every Chindogu.

Here is your guide to creating a Chindogu:

First, brainstorm problem-solving ideas and abandon any that obviously work. Then build a prototype of the best idea that looks good but on closer examination isn't. Make sure to test and verify that the prototype indeed wasn't worth the effort. Finally, reward yourself for successfully producing an almost useful creation!

So what is the point of Chindogu? Unbridled curiosity and creation, of course! Problem-solving requires new approaches and different perspectives. What better way than untethering from the practical to explore the vast depth of what's imaginable in order to create something entirely new, albeit if not quite entirely useful.

Josh is a former space nerd turned tech nerd who founded Nerd Nite Orlando on Pi Day 2013. In the second grade he entered an invention contest for his "fishing backpack" that was just his school backpack plus car air freshener. He didn't win any prizes, not even an honorable mention.

NERD NITE SAN DIEGO

CASKETS, CORPSES, AND BIERS, OH MY!: A Brief Look at Death Care and the History and Science of Embalming in the US

by Deanne M. Rugani

Let's discuss death care, embalming, and what happens to your body after you die! I graduated from SF College of Mortuary Science in 2002 with a degree in mortuary science. I hold an embalmer's and funeral director's license in the state of California. I've worked in death care for 21 years and I'm currently employed as a forensic autopsy specialist at the San Diego County Medical Examiner, so you could say I know a little about death.

Before I dive in, there are a few misconceptions about death that I should clear up.

1. **A dead body can sit up on its own:** When someone tells you about the time they were in a mortuary and saw a body sit up, they're fibbing. Have you ever tried to sit up without using any muscles? Go ahead and try . . . I'll wait. Couldn't do it, right? Yeah, neither can a dead body.

2. **Embalming will preserve someone forever:** Embalming chemicals once contained ingredients like arsenic and mercury that could keep a decedent preserved for an extended period of time. Today's chemicals are designed to disinfect, sanitize, and temporarily preserve long enough to get through services. Some decedents can

maintain their funeral-day state for years with modern chemicals, but that's not what they're designed for.

3. **Time of death can be determined by liver temp:** Baaahahaha-hahaha! You watch too much *CSI*.

4. **Decomposition is a process in which worms "eat you":** No. I mean, yes, worms can eat you if you die outdoors and lie there long enough, but decomposition is different—your body essentially eats itself—I'll discuss more later.

5. **Hair and nails continue to grow after death:** Skin will dehydrate and recede to give the appearance of hair and nails growing, but I promise you, Uncle Jack did not grow that handsome five-o'clock shadow after he died.

As previously mentioned, I'm a licensed embalmer / funeral director. If you don't know what those are, you're not alone. An embalmer is a person who disinfects, preserves, and restores a dead human body to a natural, life-like appearance using preservative chemicals. Some embalmers are also restoration artists and do reconstruction, cosmetics, and hairstyling. A funeral director is a licensed professional who specializes in all aspects of funeral service, including arranging/directing funeral services and providing support to the bereaved. Other common titles are mortician or undertaker.

> *Death—noun \déth\: The extinction of life. A permanent cessation of all vital functions without capability of resuscitation.*

Everything that happens immediately prior to and in the hours following death is important to an embalmer and can affect the embalming process and overall appearance of the deceased. Here's a quick overview of what happens to your body when you die.

Antemortem (Before Death)

In a natural death, the time immediately preceding death is called the agonal state. This can last a few minutes to several hours. During this state the body starts to experience changes including thermal changes, tissue moisture changes, slowing of blood flow causing coagulation/clots, and the beginning of bacterial transmigration, which is when putrefactive organisms, normally confined to the colon, migrate into the bloodstream and are transported throughout the body. These organisms begin the process of autolysis, the enzymatic digestion of cells by their own enzymes—that's decomposition, your body digesting itself. In a sudden, tragic death the agonal state would generally not occur and autolysis would begin postmortem.

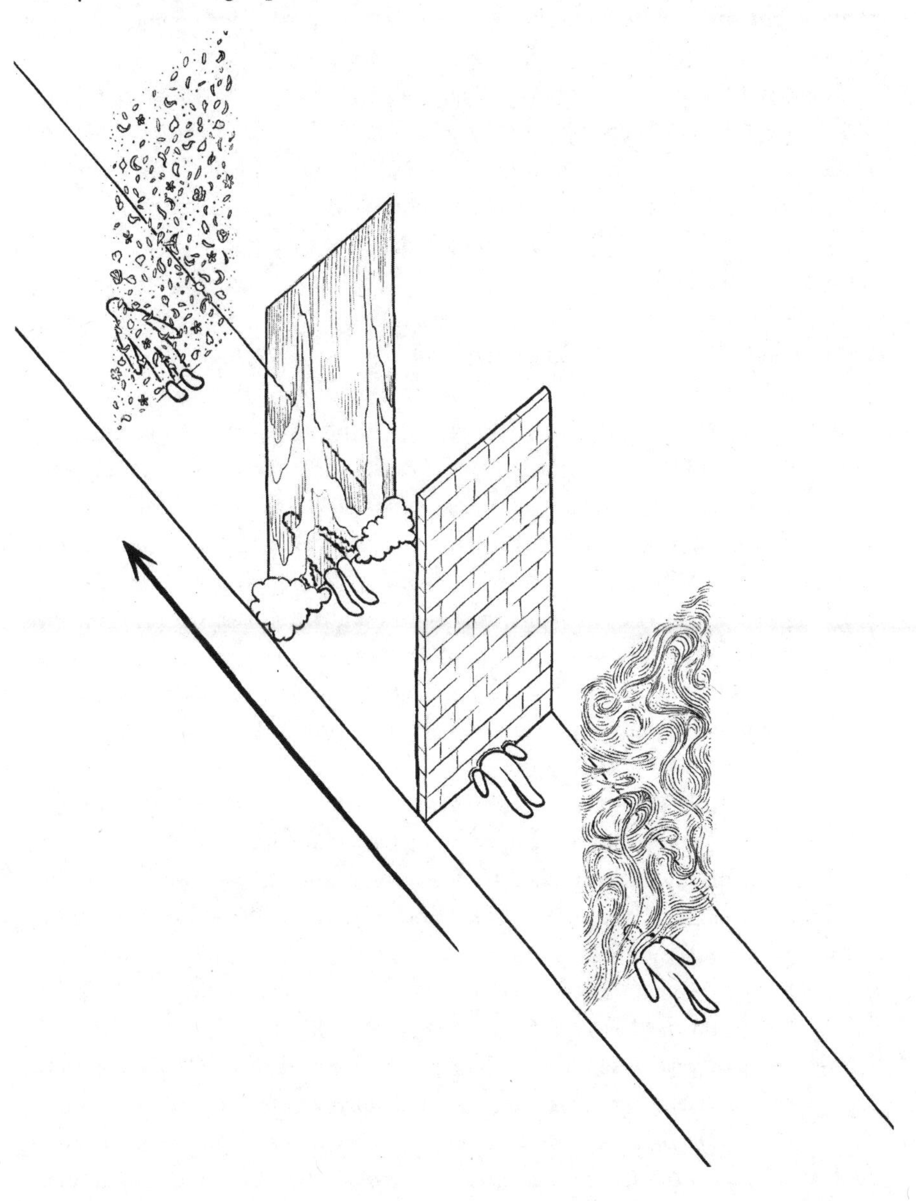

Postmortem (After Death)

The process of dying has two stages, *somatic* and *cellular*.

During somatic death you'll first experience *clinical death*. This will happen by syncope (heart death), coma (brain death), or asphyxia/apnea (lung death). Clinical death generally lasts no more than five minutes. If the right conditions permit and timely medical attention is received, a person *might* be revived during this time.

Next you'll experience *biological/irrevocable death*. By this time organs have undergone irreversible changes and restoration of life would be impossible. Basically, you're dead dead. This is the stage where the "Mortis Sisters" appear: *algor mortis* (loss of body heat), *livor mortis* (discoloration), and *rigor mortis* (stiffening of the body muscles). This usually appears eight to 20 hours postmortem and generally lasts 10 to 72 hours.

Cellular death occurs one to three hours following somatic death. This is the death of individual body cells following the depletion of oxygen and nutrients in the body.

Embalming is the process of chemically treating a dead body to reduce the presence and growth of microorganisms, to slow decomposition, and to restore an acceptable physical appearance to the body.

Modern embalming is done through arterial injection of preservative chemicals, mostly formaldehyde or glutaraldehyde, which replaces the blood in the body. These chemicals destroy organisms and enzymes already present in the tissues and cause "fixation" or "coagulation" of tissue cell proteins, which alters them in such a way that they're no longer a suitable medium or food for bacterial growth. The altered proteins have a resistance to both autolytic and bacterial enzymes, but only temporarily.

Embalming can be done from any artery in the body. Embalming fluid is injected into an artery, and blood is drained from a neighboring vein.

A pre-embalming assessment will note weight, height, age, disease, cause of death, discolorations, and overall condition of body to determine what fluid mixture will be used, what the index (percentage of formaldehyde) will be, if there's a need for specialty fluids, and how many gallons may be needed.

Once an assessment is complete the features will be set (eyes and mouth closed); the body is washed and disinfected; and rigor mortis is "broken" through massage. An injection artery is chosen (most often the right common carotid), then raised along with its accompanying vein (most often the right jugular), and both are ligated. The embalming machine is set to the embalmer's preferences. An injection tube is placed in the artery and secured, a spring forceps is placed in the vein for drainage, and the machine is turned on. During embalming the body will be washed and massaged to get blood moving and clear congested areas. Once injection is complete, instruments are removed, incisions are closed, and the thoracic and abdominal cavities are then aspirated and treated with a high-index fluid to ensure preservation. The body is washed a final time and the decedent is set in a natural, peaceful position to await services.

Thanks for attending my Dead Talk.

Deanne Rugani graduated from SF College of Mortuary Science in 2002 with a degree in mortuary science and now holds an embalmer's and funeral director's license in the state of California. Deanne has worked in death care for 21 years and is currently employed as a forensic autopsy specialist at the San Diego County Medical Examiner.

NERD NITE KANSAI

WILDLIFE DETECTIVES: The Science and Stories of "Animal CSI" in Investigating and Solving Wildlife Crime

by Dr. Rebecca N. Johnson AM

Wildlife crime is one of the most lucrative transnational crimes (along with drug, weapons, and human trafficking). It is a serious and often confounding crime that can target vulnerable species, as it drives extinction and is often closely associated with organized crime. When I studied the genetics of fruit flies and weaver ants at university in Australia, I never dreamed that one day I'd be applying my scientific skills in the field of wildlife forensics, the science that interfaces with the world of wildlife investigation and wildlife crime.

The world of wildlife crime is one that constantly surprises, but one where you feel like there is a genuine role for science and good communication of that science. This is a world where people smuggle bird eggs hidden in their pants across the world with the goal of selling them illegally for a large profit. This world has seen me applying my genetics skills to assist in cases of "squirrel smuggling" or the "weaponization" of a deadly snake.

The investigative role of wildlife forensics is also very broad. It can be working to identify a fish fillet to determine if the species fish indeed matches the label (Hint: Your tuna might not be tuna.), it can be working with the aviation industry to identify the remnants of an animal that might have collided with a plane, or it can be providing an identification of a potential pest that has been inadvertently brought into the country in an agricultural shipment. The "forensic" aspect of this work is to ensure that the processes, protocols, and results are conducted in a way that meets the expectations of the legal system. Is the sample stored securely? And is the test fit for purpose?

During my time as a visiting professor at Kyoto University Wildlife Research Center in Kyoto, I had worked on demonstrating how science is used in wildlife investigation, including some work that we did with some students in Kyoto to investigate one of Japan's most beloved foods—unagi (eel).

Freshwater eels (genus *Anguilla*) turn out to be, based on the number of eels trafficked per year, the most common victims of wildlife crime. In 2018, some 15 million eels were seized in the context of 153 arrests made. It is estimated that some 350 million eels (that is a *lot* of unagi) are trafficked from Europe each year. These juvenile eels are then raised in captivity, eventually making their way to dinner plates. So where does the forensics come in? Genetic information from confiscated eel can be used to determine its geographic origins. Think of it like a 23andMe for eels. Genetic testing in humans has revealed nefarious crimes, and the same can be said of eels.

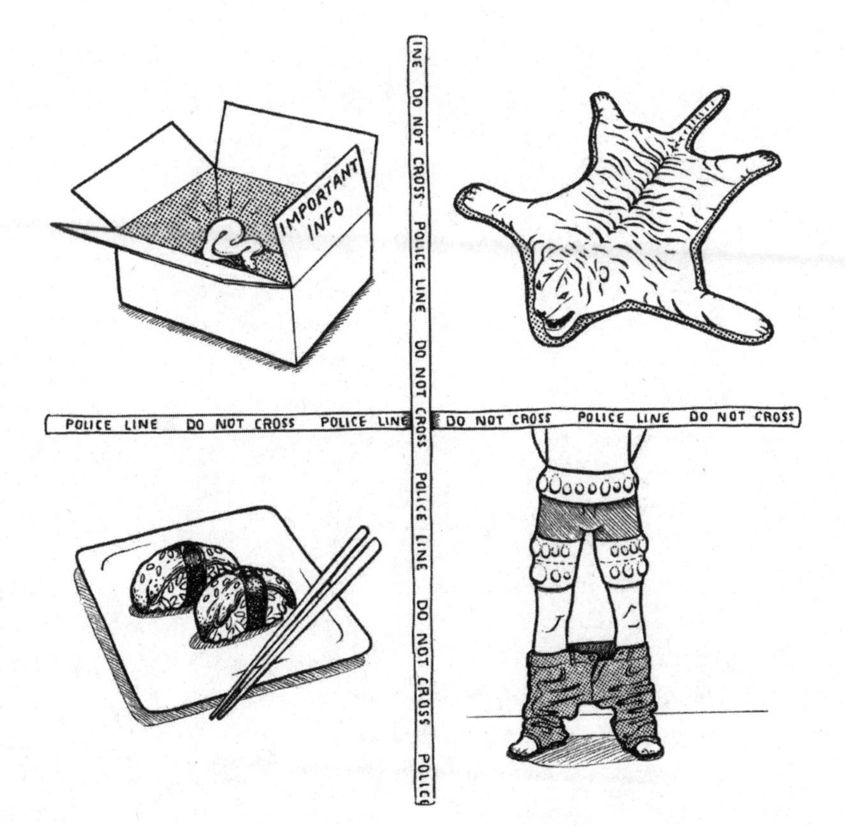

To match a trafficked eel to its ancestors around the world, we need a large database of background genetics, data that is often generated from specimens held in natural history collections around the world. Whereas most people think of natural history museums as the place you go to see old bones, behind the scenes there is a buzz of crime-solving activity, discovering unknown species the world over, and describing how these species are changing with climate change. And yes, also old bones.

Dr. Rebecca Johnson is the chief scientist and associate director for science at the Smithsonian National Museum of Natural History. Rebecca is most passionate about the importance of museum science as it relates to wildlife conservation and reducing the illegal wildlife trade. She is also dedicated to the importance of STEM, particularly women in STEM, in early and lifelong education in contributing to positive environmental outcomes.

NERD NITE DC

CUT IT OFF!:
A Civil War Amputation

by John Lustrea

What would happen if you were wounded on a Civil War battlefield? It's not a pleasant question to contemplate in the twenty-first century, is it?

First, if you're fortunate enough to suffer a clean break, you'll have a good chance of being fine, as normally with a fracture or a clean break, you put a cast on the limb and don't bang it into things, and the bone puts itself back together (now might be a good time to mention I'm a historian and not a doctor). But if a bullet strikes your bone, you might be in trouble (especially if it's your skull), because when a bullet strikes a bone, the bone will likely shatter. Today you can recover from such a wound with the help of pins, screws, and so on, but in the Civil War, that was beyond the doctors' skill to heal. They therefore turned to amputations, as they were the most common form of surgery in the Civil War—with over 60,000 conducted!

So what did the most common Civil War amputation surgery look like? I thought you'd never ask!

If the patient didn't already come to the operating table with a tourniquet on, the surgeons would apply one so blood didn't shoot everywhere once the cutting began. Though, if done well, an amputation shouldn't be all that bloody.

Next, the surgeons would administer anesthesia, usually either a bit of ether or the less explosive (and therefore preferred) chloroform. Believe it or not, even in the 1860s, anesthesia was used in more than 95 percent of Civil War surgeries.

With the patient safely asleep (for only fifteen minutes), the surgeons broke the skin with a scalpel—incidentally, the scalpel has barely changed since the Civil War and is still used by doctors around the world every day. With the skin cut, the doctors would roll it back like a shirtsleeve to use later as a skin flap for the stump. Next, they used a knife to cut through the soft tissue (muscles, ligaments, and tendons), which at that point was all that held the limb to the bone.

Enter the bone saw! Once that very self-explanatory implement was through working, the limb was tossed elsewhere to be buried.

The next few tools used in the procedure reveal just how precise a Civil War surgeon's tool kit was. I'm guessing precise is not what comes to mind when thinking of Civil War surgeons?

They began the second half of the operation with something called a tenaculum. The purpose of this hooked tool was to locate all the veins and arteries in the stump to ensure they were properly tied off so blood could continue to circulate properly. No sense in having the patient bleed out after the operation.

Next was the bone file. As with a fingernail, we don't want any sharp bits pointing out of the stump causing discomfort, so the surgeons would file down any rough edges. They would then brush away any hard bone dust from the area, which would hurt if sewn up in the stump.

Finally, the surgeon would finish the procedure by rolling the skin back down over the stump and sewing the patient up with needle and thread.

How likely was someone to survive an amputation? It depends.

The farther away from the core of the body, the more likely someone was to survive, which makes sense to my non-medical mind. The mortality rates would range from a 3 percent for a finger to 83 percent for an amputation of the leg at the hip. Averaged out, there was a 26 percent mortality rate, which sounds just okay, though is much better when compared with the mortality rate of civilian amputees before the Civil War, which was 50 percent. Civil War soldiers were nearly twice as likely to survive an amputation as civilians due to a number of factors, but foremost among them was speed. The closer to the time of injury an operation was performed, the more likely a patient was to survive. Civilians often waited to go to a hospital as a last resort, but soldiers who had their operations sooner (thanks to the newly formed Ambulance Corps) fared much better.

The elephant in the room is infection, which was nearly universal since the doctors didn't know about germ theory and thus rarely washed their hands between patients. Infection was so common that doctors often thought pus was simply a part of the healing process (look up *laudable pus* for a fun time). Because

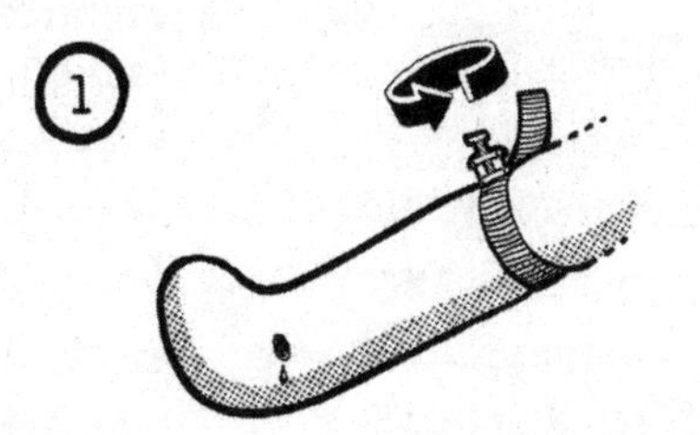
1

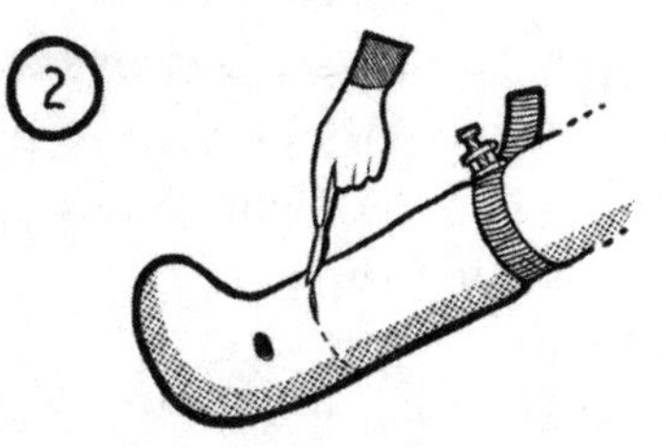
2

3

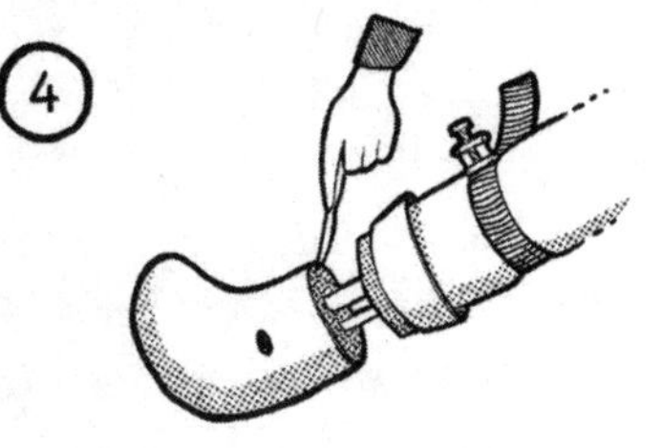
4

5

6

7

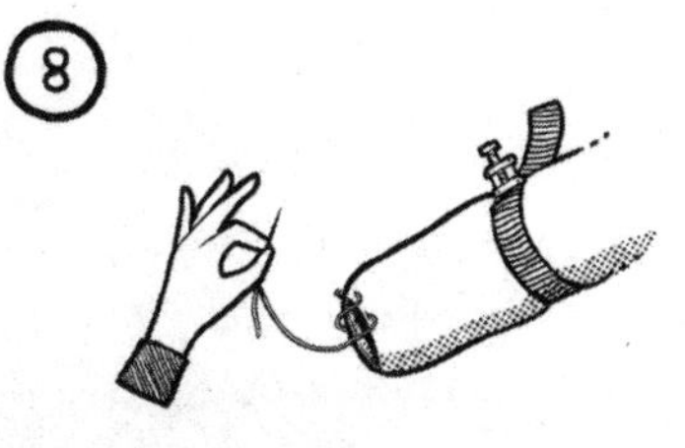
8

9

Civil War surgeons didn't know what infection was, when patients died from it, they were listed as a statistic under whatever relevant symptom they suffered from, rather than the actual cause of death.

The biggest difference between amputations then and now is not the actual nuts and bolts of the operation itself, but how rare it is to get an infection in the operating room today.

So there you have it. You now know how to amputate a limb . . . more or less. You might not be a doctor, but you are now equipped to play one on TV.

John Lustrea is the director of education at the National Museum of Civil War Medicine. He earned his Master's degree in Public History from the University of South Carolina and has previously worked at Harpers Ferry National Historical Park.

NERD NITE ALPINE, TX

FERMENTATION:
A Cultural Story

by Amy Oxenham

Nearly four thousand years ago, you could be sentenced to death for pouring a short pour of beer. Yep, fermentation was as important millennia ago as it is today.

The Merriam-Webster Dictionary defines *fermentation* as "an enzymatically controlled anaerobic breakdown of an energy-rich compound (as a carbohydrate to carbon dioxide and alcohol or to an organic acid); broadly, an enzymatically controlled transformation of an organic compound, or the production of energy in the absence of oxygen." But in simpler terms, it's nature's way of breaking down nutrients and sending them back into the environment to become the raw ingredients for everything we see around us.

Fermentation is a metabolic process of microorganisms, especially, but not limited to, bacteria and fungi, converting carbohydrates into material to be used for energy and protein synthesis. It's a fundamental part of every living organism and basic to our very cells. Not only is it a powerhouse for freeing energy and materials, this process is also a significant way of preserving and enhancing food products for human consumption. In fact, humans have been harnessing the power of microbes before we could even understand the mechanisms by which fermentation worked, as fermenting food and beverage is a practice that most peoples across the globe all arrived at regardless of time or geography. It can be done safely with very few tools and ingredients and can also happen spontaneously at times without human assistance, therefore

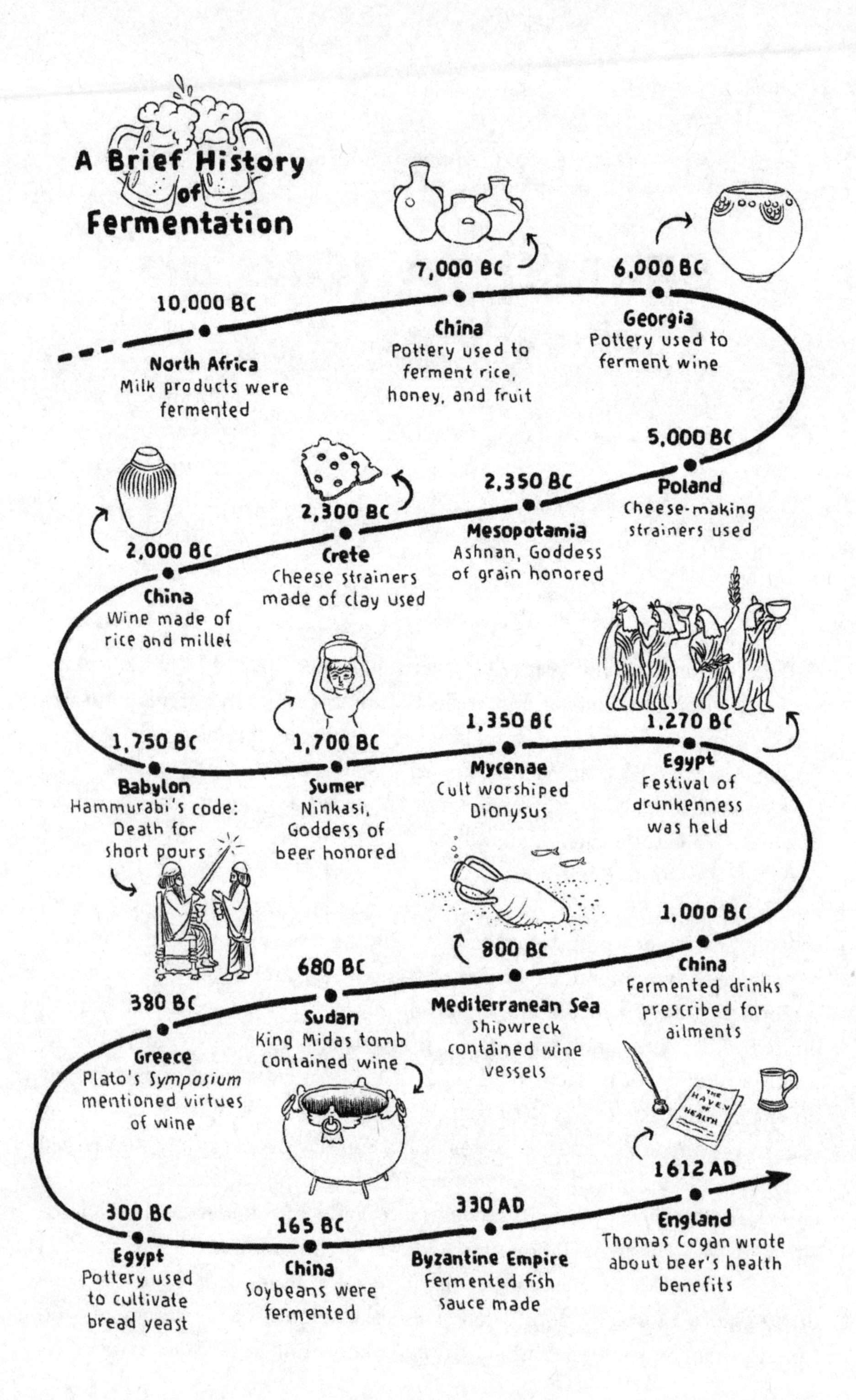
A Brief History of Fermentation
10,000 BC
North Africa
Milk products were fermented
7,000 BC
China
Pottery used to ferment rice, honey, and fruit
6,000 BC
Georgia
Pottery used to ferment wine
5,000 BC
Poland
Cheese-making strainers used
2,350 BC
Mesopotamia
Ashnan, Goddess of grain honored
2,300 BC
Crete
Cheese strainers made of clay used
2,000 BC
China
Wine made of rice and millet
1,750 BC
Babylon
Hammurabi's code: Death for short pours
1,700 BC
Sumer
Ninkasi, Goddess of beer honored
1,350 BC
Mycenae
Cult worshiped Dionysus
1,270 BC
Egypt
Festival of drunkenness was held
1,000 BC
China
Fermented drinks prescribed for ailments
800 BC
Mediterranean Sea
Shipwreck contained wine vessels
680 BC
Sudan
King Midas tomb contained wine
380 BC
Greece
Plato's Symposium mentioned virtues of wine
300 BC
Egypt
Pottery used to cultivate bread yeast
165 BC
China
Soybeans were fermented
330 AD
Byzantine Empire
Fermented fish sauce made
1612 AD
England
Thomas Cogan wrote about beer's health benefits

making it relatively low-tech and accessible to a range of people across all backgrounds all around the world.

Human culture took a sharp turn once we figured out how to harness the power of fermentation; it has yielded some of mankind's greatest culinary creations, including such highlights as bread, coffee, chocolate, tofu, cheese, and beer, of course! Fermentation practices encouraged our first vital steps into agriculture and ensured that humans could be nourished in some of the most inhospitable conditions as our species spread out from the cradle of civilization. Fermentation even played an integral part in enabling our ancient ancestors to leave behind our hunter-gatherer ways and establish agriculture and communities. In *Tasting Beer*, Randy Mosher explains, "The story begins around 10,000 BC, just after the glaciers of the last Ice Age retreat to the north. As they do, the vacated land becomes grassland. The Neolithic people in the hill country of what is now Kurdistan start to use the grasses as a good source of nutrition, saving the best seeds . . . this is the beginning of agriculture and purposeful fermentation."

Some of the first evidence of fermentation practices among people:

- 7000–6600 BC: Pottery from a Neolithic community in China's Yellow River Valley is chemically analyzed and determined to have contained a fermented beverage of rice, honey, and fruit.
- 6000 BC: Pottery from a Neolithic community in Georgia is determined to have held wine (and even the remains of a fruit fly).
- 5000 BC: What are believed to be strainers for making cheese are found in Poland; upon analyzing it is determined that there are milk residues.

Fermentation in the lives of ancient people around the world:

- 10,000 BC: Fermentation of milk products in North Africa.
- 2350 BC: In Mesopotamia, Ashnan was goddess of grain.
- 2300 BC: Cheese strainers made from clay have been found on Crete.
- 1750 BC: Hammurabi's Code includes punishment by death for short pours of beer.
- 1700 BC: Ninkasi is the Sumerian goddess of beer.

- 1350 BC: There is believed to be a cult to Dionysus in Mycenean community.
- 1270 BC: During the rule of Seti I in Egypt, the Festival of Drunkenness, taken from mythological stories of how beer saved the world, is celebrated annually.
- 800 BC: In modern Sudan, Bes is a god often depicted drinking beer.
- 800 BC: Shipwrecks in the Mediterranean Sea (from modern Lebanon and Syria) contained vessels full of wine for trade.
- 680 BC: Drinking vessels in the supposed tomb of King Midas are determined to contain a fermented beverage of grapes and assorted herbs.
- 385–370 BC: Plato references the virtues of wine in the *Symposium*.
- Also during this period, Aristotle refers to wine making in many of his writings.
- 300 BC: Evidence discovered in clay jars in Egypt suggests yeast cultivation for the express purpose of bread making.
- AD 330: The Byzantine Empire develops recipes for a fermented fish sauce called garum.

Fermentation in the ancient Far East:

- 7000 BC: It's been determined through chemical analysis of clay pottery that the Neolithic peoples of China drank fermented beverages.
- 2000 BC: During the Shang and Zhou dynasties, bronze vessels held wine made from rice and millet.
- It is also during this time that ancient Chinese medicine focused on dietetics as therapy. It is believed that fermented beverages held an important place in medicine and treatment.
- Eleventh century BC: The Zhou dynasty placed a huge emphasis on medicine and on dietetics as treatment. Fermented beverages of rice and fermented foods of black soy were prescribed as a carrier for healing herbs for a variety of ailments.
- 165 BC: There is evidence of fermented soybeans in ancient China.

The view of fermentation in the ancient world:

- Fermentation was viewed as a process largely controlled by god or goddesses, much like agriculture. This is reflected in the long list of deities from around the globe responsible for harvest, wine, beer, fermentation, and even drunkenness. At least, that is what the historical and archaeological evidence shows.
- It is not until later years (300 BC approximately) that ancient people begin to focus on diet and nutrition as medicine and include fermented beverages especially in treatment and prevention of disease.

Humoral medicine and fermentation:

- Influenced by Hippocratic doctrine, humoral medicine indicates that certain humors considered to be out of balance can be treated successfully with mixtures of healing herbs.
- Healing beers with elemental qualities are also prescribed for treating humors, an idea that persisted to modern time as well:
    - "Seeing for the most part there is felt no small bitternesse in Beere, there is no doubt but all Beere is hot: and how much more bitter it is, the hotter it is. But notwithstanding, I thinke hopes in Beere maketh it colder in operation, because it purgeth choler."—Thomas Cogan, *The Haven of Health*, 1612
    - In a brewing text from 1633, beer is reported to "make good blood and expelleth phlegmatic and melancholy humors."

Fermentation is certainly a cornerstone of human civilization. The practice of fermenting has not only created culinary delights but also enriched our species' culture (pun intended) in innumerable ways!

Amy Oxenham is a fermentation enthusiast and professional brewer. Amy's dedication to fermented food products started with farming and an interest in food preservation techniques. Since those early days, Amy has committed her education and career to fermentation, becoming a professional brewer in 2013, earning a BS in Biology, completing a Certificate in Brewing Technology, and now currently working on a graduate degree in brewing science and operations. Amy loves all things fermentation!

NERD NITE MADISON

FIRE: Of Flames and Friendship

by Lee M. Bishop, PhD

I was a bearded and disheveled chemistry PhD candidate, and I thought I had everything (with the glaring exception of my life) figured out. I had an atomic explanation for everything. My tie-dye T-shirts are long, twisted, tangled cellulose chains. The best tie-dyes do not rinse away over time because they attach themselves through carbon-oxygen bonds to cellulose chains. The yellow color of pee comes from degraded hemoglobin molecules, which are what make our blood red. Stuff like that. What else did a person need to know? Everything was made of atoms, and you could understand all about it by its nice, tidy, atomic structure. At least that's what I thought until one fateful day. But first . . .

In those days I spent a lot of time in the lab, and not a lot of time showering or trimming my beard. However, once, in a brief glimmer of the life that could be, I went camping with some friends. As any irresponsible chemist would do, I brought along methanol spray bottles loaded with lithium and copper chloride salts with which I could produce pink and green fireballs. This of course sparked questions about fire that I was more than happy to expound upon. What are ashes? you ask. Well, they are made of substances called metal oxides. Did you know they can be mixed with water and fat to make soap? In fact, this is probably how soap was discovered, as people rubbed ashes and water on their grease-coated metal pots and noticed suds. Why do pine needles go crazy on a fire? you ask. Well, they contain lots of a substance called pinene, which burns more quickly than the molecules in wood. Not enough of an explanation for you? Well okay. Pinene burns more quickly than wood because pinene molecules are much smaller than the molecules in wood, which means pinene can vaporize, mix with oxygen, and burn more quickly.

Everything was going fine until this next question. What are flames made of? you ask. Um . . . Um . . . I remember it so vividly. I was sitting on a log and didn't know the answer to the question. You might say I was stumped. Zing!

A lot of things crumbled to ashes for me at this point. Not only were my atomic lenses unable to peer into flame itself, but something else was wrong. Why was it so important that I know the answer to this question? Surely a large part of it was curiosity and being able to feel a sense of mastery over the workings of the world. But eventually I realized that I felt like I was letting my friends down. They came to me for entertainment and knowledge, right? That is why they liked me, right?

Well, I will not let you down, intrepid reader. Fast-forward through the camping trip, the return to the lab, and poring over research articles. Put on your atomic-scale thinking caps.

Let's zoom in and imagine we are wax molecules on their final fateful journey through the eye of a candle flame. We are made of long chains of carbon atoms. As the flame above us roars, we absorb its energy and our atoms begin to jiggle ever faster, eventually jiggling so much that we begin to move around one another. That movement marks our transition to liquid wax, which is then drawn into the flame wick. We continue to absorb energy as we move, nanometer by nanometer, up the wick and closer to the flame. Our atomic chains begin to dance so wildly that we bounce off the wick and enter the gas phase. In the final moments of our existence, our long atomic chain bodies have absorbed so much energy that they fly apart and send fragments careening into a demolition derby of atoms and molecule fragments. This atomic jumble includes oxygen atoms, with which our carbon and hydrogen atoms begin to bond. It's this insane mixture of highly energetic molecular fragments that gives off the flame's blue light through a process called chemiluminescence. Just as happens in glow sticks and the butts of lightning bugs, highly energetic electrons produced through violent atomic rearrangements release their pent-up energy in the form of light.

As they continue to move outward from the center of the flame, many of the atomic remnants of our bodies fully combine with oxygen atoms to form carbon dioxide and water, and the atomic dance is done. But others aren't so lucky. When the wax fragments can't find oxygen atoms, they rearrange themselves into jumbles of mostly carbon atoms known as soot. While soot is black at room temperature, in the fiery belly of a candle flame, soot molecules glow red-hot, emitting the red-orange color we know and love. This is known as incandescence, just like in old-school lightbulbs.

So there we have it. Flames are not just wisps of pure light. That light comes from atoms, and those atoms are part of molecules that are too energetic or short-lived to be bottled like cotton or dyes or pee. I feel better knowing the answer, don't you?

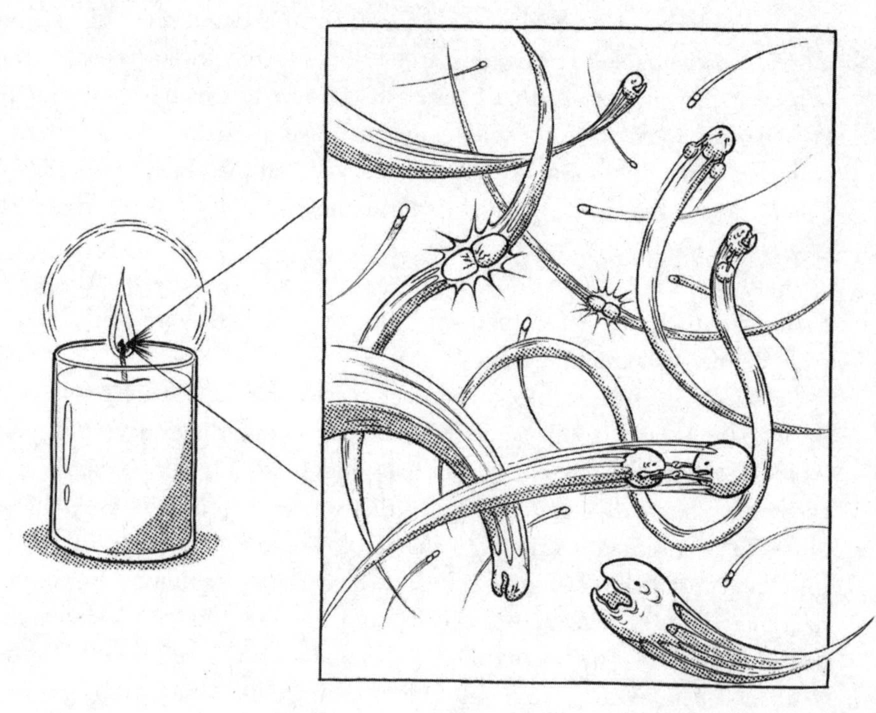

Fast-forward a few more years. I now bathe regularly. Okay, semi-regularly. I also don't worry as much about not having a tidy explanation for everything, and I have discovered that people still want to go camping with me. Maybe explaining fire wasn't the most important part of friendship? I also have two young kids now, so I don't have much time for poring over research articles to sate my curiosity. However, my son just asked me why poop is brown . . .

Lee Bishop is a chemist and educational experience designer based in Berkeley, California. He once wrote a blog called Science Minus Details Plus Weed, *which he changed to* Science Minus Details *to be more professional.*

OUR BELOVED NERD NITE BOSSES

What does it take to be a Nerd Nite boss? Technically, you must have attended at least one Nite in person to understand our event format, spirit, and balance of formality/informality. Unofficially, you need to be organized, curious, and enthusiastic. A little passion goes a long way, and a lot of passion goes even farther. Nerd Nite bosses are champions of the curious and the silly and are willing to toil behind the scenes, engaging with presenters, venues, occasional sponsors, social media, local media, and finicky A/V to ensure they're delivering fun-yet-informative events. All Nerd Nite bosses and co-bosses around the world do this as a labor of love. Most don't earn any money at all, and those that do make ones of dollars/euros/yen/you-get-the-picture per month as they keep the shows going for everyone's enjoyment. Some bosses are brand-new. Some are retired. Some closed up shop during the pandemic. And some have been leading their Nites for more than a decade. These are the folks you should thank!:

Aachen: Christine Kathrein
Aachen: Mai-Thi Nguyen-Kim
Abilene, KS: Amy Feigley
Albany: Lindsay Bianco
Albany: Pascelle Saint Laurent
Amherst, MA: Ann Tweady
Amherst, MA: Josh Rousseau
Amsterdam: Abigail Stevens
Amsterdam: Alfonso
Amsterdam: Anne de Beurs
Amsterdam: Charlaine Roth
Amsterdam: Dirk Boonzajer Flaes
Amsterdam: Huyen
Amsterdam: Johannes Oberreuter
Amsterdam: Laura Sanna
Amsterdam: Lisa Hartgring
Amsterdam: Lucas Ellerbroek
Amsterdam: Marieke van Doesburgh
Amsterdam: Ning Lin
Amsterdam: Sandrine Haene
Amsterdam: Shanna Haaker

Amsterdam: Stella
Amsterdam: Whitney Pattinaja
Anchorage: Chris Linn
Anchorage: Rachel Mills
Ann Arbor: Amber Conville
Ann Arbor: Elyse Aurbach
Ann Arbor: Hadley Lord
Ann Arbor: Liz Lamoste
Ann Arbor: Mariah Cherem
Ann Arbor: Sara Wedell
Ann Arbor: Emily Murphy
Appleton: Nate Ansari
Asheville: Ashley Hart
Asheville: Ve Magni
Atlanta: Andy Fisher
Atlanta: Matt Young
Atlanta: Tracy Galasso
Auckland: Beau Pontre
Auckland: Ben Curran
Auckland: Jimmy Dalton
Auckland: Lisa Strover
Auckland: Tim McNamara
Augsburg: Jasmine Mahr
Austin: Amy Cavender
Austin: Cameron Russell
Austin: Erin Parr
Austin: Jacob Weiss
Austin: JC Dwyer
Austin: Kain Shin
Austin: Lewis Weil*
Austin: Mickey Delp
Baltimore: Jaye Ferrone
Baltimore: Jen Sahm
Baltimore: Joel Green
Baltimore: Patrice Woodard
Baltimore: Roy Prouty
Banff: Amanda Sarka
Banff: Lisa Belanger
Barcelona: Christine Hart
Berlin: Bettina Rech
Berlin: Jan Renz
Bethlehem, PA: Glen Tickle
Big Bend, TX: Betsy Evans
Big Bend, TX: Joslyn Sandlin
Big Bend, TX: Victoria Contreras
Bloomington, IN: Leanne Zdravecky
Bloomington, IN: Matt Neer
Boca Raton: Jennifer Fierman
Boca Raton: Maria Uglum
Boca Raton: Matt Hogan
Boca Raton: Troy Bernier
Boise: Matt Laye
Bonn: Jessica Kraus
Bonn: Paul von Golaszewski
Boston: Chris Balakrishnan
Boston: Jeremy Kay
Boston: Mary Lewey
Boston: Tim Sullivan
Boulder: Alison Gilchrist
Boulder: Anne Pierce
Boulder: Benjamin Nault
Boulder: Julie Sadino
Boulder: Matt Hartsock
Boulder: Max Boykoff
Boulder: Raquel Salvador Gallego
Bozeman: Jill Mellecker
Bratislava: Robert Srnka
Braunschweig: Britta Eisenbrath
Brighton: Anna Downie
Brighton: Kate Doran
Brighton: Katy Howland
Brighton: Partha Das
Brownsville: Al Alder
Brownsville: Catheline Froehlich
Brownsville: Joey Key
Brownsville: Max Abrahamson
Brussels: Michiel Buydaert
Buenos Aires: Ana Bedacarratz
Calgary: Aaron Noel
Calgary: Erika Smith
Calgary: Frank Koutis
Calgary: Joey Windsor
Calgary: Krysta Siever
Cambridge, UK: Amber Kaplan
Camden, Australia: Jessica Bruce
Camden, Australia: Andreas Mertin
Chapel Hill: Susan Brown
Chattanooga: Devori Kimbro
Chattanooga: Justin Vizaro
Chicago: Dan Rumney
Chicago: Jason St. John
Chicago: Jessica Schmidt
Chicago: Laura Lanford
Chicago: Rebecca Anderson
Christchurch: Glen Tregurtha
Christchurch: Nicole O'Hearne
Cincinnati: Cat Musgrove
Cincinnati: Stephen Morro
Cologne: Florian Liss
Cologne: Marina Bommas
Colorado Springs: Flip Aguilera
Colorado Springs: Mari Daman-Carles
Columbia, MO: Sam Stoeckl
Columbia, MO: Sarah Elizabeth Stoeckl
Columbus: Atom Vincent
Concepción, Chile: Hector Enriquez Diaz
Copenhagen: Ana Verissimo
Copenhagen: Halle Fabrin Hansen
Cork: Matthew Blair
Coventry, UK: Gareth Jenkins
Dallas: Dan Hoffman
Davis: Kevin Wan
Denton: Ali Kelly
Denton: Annette Becker
Denton: Eric Robinson
Denton: Matt Henry
Denton: Matt Kernan
Denton: Shaun Treat
Denver: Ana Draghici
Denver: Earl Anema
Denver: Hanna Aucoin
Denver: Ian Kimsey
Denver: Jon Mohr
Denver: Sara Wilson
Des Moines: Grace Vaziri
Des Moines: Jim Adelman
Detroit: Alex Gjerovski
Detroit: Amber Conville
Detroit: Anna Van Toia
Detroit: Bubba Ayoub
Detroit: Elissa Zimmer
Detroit: Jenny LaDuke
Detroit: Jensi Simkins
Detroit: Josh Diskin
Detroit: Liz Lamoste
Detroit: Maggie McGuire
Detroit: Nathan Hughes
Dresden: Marcus Wermuth
Dublin: Ashwini Deshpande
Dublin: Karl Schakermann
Duluth: Adam Brisk
Duluth: Crystal Pelkey
Durham: Courtney Fitzpatrick
Durham: Jeremy Kay
Dusseldorf: Florian Liss

East Bay: Ian Davis
East Bay: Rebecca Cohen
East Bay: Rick Karnesky
East Bay: Sarah Houghton
East Bay: Scott Weitze
Eau Claire: Kelsey Deetz
Edmonton: Adam Rozenhart
Edmonton: Kati Kovacs
Edmonton: Lauren Albrecht
Edmonton: Lisa Belanger
Edmonton: Marc-Julien Objois
Edmonton: Rebecca Fletcher-Calder
Edmonton: Ross Lockwood
Edmonton: Tammy Bearht
Emerald Isle: Tatsiana Bylund
Erlangen: Andreas Tobola
Erlangen: Christian Sauter
Fairbanks: Jessica Johnson
Fairbanks: Jon Celmer
Fargo: Dan Novacek
Fargo: Darby Miller
Fargo: Erik Moran
Fargo: Hope Novacek
Fargo: Karen Glover
Fargo: Tracy Kurtz
Fargo: Zachary Duval
Fayetteville, NC: Darby Ann Miller
Fayetteville, NC: Owen Williams
Fort Collins: Dan Bonomo
Fort Collins: Justin Fritz
Fort Collins: Senne Van Loon
Fort Collins: Terence Paige
Fort Lauderdale: Deven Blackburn
Fort Lauderdale: Gary Bremen
Fredericksburg: Nancy Stalik
Fredericksburg: Sandra Fedowitz
Freiburg: Caterina Mesina
Freiburg: Karen Zamb
Freiburg: Lilli Schaefer
Freiburg: Thea Glaser
Fresno: Jennifer Ward
Friedrichshafen: Nicolai Mueller
Gainesville: Chris DiScenza
Gainesville: Greg Webster
Gainesville: Joanna Karavolias
Geneva: Barbara Sandoval
Geneva: Indrit Sinanaj
Geneva: James Hamilton
Grand Rapids: Adam Hyde
Graz: Xiaoao Dong
Greenville, NC: Chequita Brooks
Greenville, NC: Jessica Cringan
Greenville, NC: Karen Ann Litwa
Greenville, NC: Sam Parrish
Guam: Cyrus Luhr
Halifax: Andrew Wright
Halifax: Leslie Walsh
Halle: Babette Richter
Hamburg: Falko Brinkmann
Hamburg: Julia Offe
Hamburg: Volkan Serce
Hamilton, ON: Mark Miller
Harare: Drew Rawlins
Harare: Tendai Gadzikwa
Harbor Beach, MI: Justin Schnettler
Harbor Beach, MI: Paula Mausolf
Hartford: Chris Hall
Hildesheim: Mathias Mertens
Homer: Abigail Clapp
Hong Kong: Angel Perucho
Hong Kong: Shawn Tan Zheng Kai
Hong Kong: Smaranda Badea
Honolulu: Becky Briggs
Honolulu: Chloe Heiniemi
Honolulu: Chris Jury
Honolulu: Christina Comfort
Honolulu: Emily Norton
Honolulu: Leon Tran
Honolulu: Lilly Buchholz
Honolulu: Lydia Baker
Honolulu: Marisa McDonald
Honolulu: Sara Wood
Honolulu: Tony Smith
Houston: Aaron Dunn
Houston: Amado Guloy
Houston: Fayza Elmostehi
Houston: Heather Ringman
Houston: Jake Joiner
Houston: Jon Martensen
Houston: Lorena Maili
Houston: Maitri Erwin
Idaho Falls: Airica Staley
Indianapolis: Carlos Peredo
Indianapolis: Doria Lynch
Ithaca: Michelle White
Jackson, TN: Rachel Guyer
Johannesburg: Charles Henderson
Johannesburg: Lereece Rose
Kansai: Aga Czeszumska
Kansai: Hiro-sato Matuura
Kansai: Katie Hill
Kansai: Yuri Minoura
Kansas City: Alison Heryer
Kansas City: Charles Huette
Kansas City: Jason Kovac
Kansas City: John Helling
Kansas City: Matthew Long-Middleton
Kansas City: Shelly McNerney
Kitchener: Becky Verdun
Kitchener: Charlotte Armstrong
Kitchener: Eric Moon
Kitchener: Lakyn Barton
Kitchener: Ryan Consell
Kyushu: Raymond Terhune
Lancaster, PA: Jay Parrish
Las Vegas: Krissi Reeves
Las Vegas: Mike Henry
Lawrence: Abby Olcese
Lawrence: Adrian Jacobs
Lawrence: Amy Schweppe
Lawrence: Becky Harpstrite
Lawrence: Chad O'Bryhim
Lawrence: Dave Trimbach
Lawrence: Elliot Pollack
Lawrence: Emily Fekete
Lawrence: Jason Keezer
Lawrence: Kate Gramlich
Lawrence: Kate Meyer
Lawrence: Kevin Liu
Lawrence: Kristin Colahan-Sederstrom
Lawrence: Pat Trouba
Lawrence: Peter Lyrene
Lawrence: Sally Chang
Lawrence: Satya Na
Lawrence: Travis Weller
Leipzig: Claudia Wagner
Leipzig: Philipp Hertel
Leipzig: Sebastian Hupfer
Lincoln, NE: Cassi Smith
Lincoln, NE: Crystal Uminski
Lincoln, NE: Emily Hudson
Lincoln, NE: Kate Hanley Schofield

Lincoln, NE: Matt Wilkins
Lincoln, NE: Rachael Disciullo
Lincoln, NE: Stephanie Berg
Lincoln, NE: Tyler Corey
Lincoln, UK: Ian Widdows
Lincoln, UK: Kate Bergens
Lincoln, UK: Shaun Sellars
Livingston, MT: Michael DeChellis
Livingston, MT: Michal DeChellis
London: Dominique Morneau
London: Ewen Callaway
London: Isabel Inman
London: Ivana Kottasova
London: Jo Wilmot
London: Kara Signer
London: Louise Inman
London: Matthew Spivack
London: Nick Perry
London: Robert Williams
London: Suzi Price
Los Angeles: Amy Lam
Los Angeles: Cyndi Lynott
Los Angeles: Erica Li
Los Angeles: Heath Rumble
Los Angeles: Jessica De Vita
Los Angeles: Katie Kaniewski
Los Angeles: Liz Janss
Los Angeles: Nicholas Peters
Los Angeles: Riley Gibbs
Los Angeles: Sean McDonald
Louisville: Bentley Mcbentleson (RIP!)
Louisville: Eileen Street
Madison: Ally Herro
Madison: Elena Spitzer
Madison: Haley Briel
Madison: Jamie Holzhuter
Madison: Julie Collins
Madison: Laura Detert
Madison: Lee Bishop
Magdeburg: Anna Thunen
Magdeburg: Constantin Kwiatkowski
Magdeburg: Heiko Weichelt
Magdeburg: Jessica Bosch
Magdeburg: Kristin Held
Magdeburg: Martin Hess
Magdeburg: Norman Lang
Magdeburg: Sebastian Bannasch
Manchester UK: Leo Gutkowski
Maryville: Cyd Hamilton
Melbourne: Allison Irvin
Melbourne: Cecily Clarke
Melbourne: Dejan Jotanovic
Melbourne: Doris Toh
Melbourne: Isabell Kiral-Kornek
Melbourne: Jack Dunstan
Melbourne: Jacobien Carstens
Melbourne: Mark Richardson
Melbourne: Steph Met
Melbourne: Steven Manos
Melbourne: Wade Kelly
Memphis: James Weakley
Memphis: Jim Adelman
Memphis: Josh Wolfe
Memphis: Stephanie Madden
Miami: Gary Bremen
Miami: Laura Chaibongsai
Miami: Marc Kaplan
Miami: Melissa Blundell Osario
Miami: Nathan Laxague
Miami: Rebecca Peterson
Miami: Shane Smith
Milan: Lauren Plavisch
Milton: Kim Fowler
Milwaukee: Crysta Jarczynski
Milwaukee: Laura Detert
Milwaukee: Matthew Staab
Minneapolis: Carissa Getty
Minneapolis: Lesley Kadish
Minneapolis: Rosae Corral
Minneapolis: Sarah Nerison
Mobile: Jennifer Stallings
Monrovia, Liberia: Ben Morgan
Monrovia, Liberia: Dara Lipton
Monrovia, Liberia: Matt Mirecki
Monterey, CA: Sharon Gavin
Montreal: David Trossman
Montreal: Keith Kolder
Montreal: Nabiha Yahiaoui
Montreal: Swathi Meenakshi
Montreal: Yuting Zhang
Munchen: Patrick Gruban
Napoli: Francesca Cagnoni
Napoli: Francesca Santoro
Napoli: Paola Scognamiglio
Nashville: Martha Girdler
Nashville: Scott Kerr
New Delhi: Abdulrahman Alhalawani
New Orleans: Champ Superstar
New York City: Matt Wasowski
Nicaragua: Adolfo Gonzalez
Nicaragua: Kristopher Mendoza
North Bay: Sarah Houghton
North Bay: Scott Weitze
North Sydney: Jade Fardouly
North Vancouver: Charlie Cook
North Vancouver: Crystal Baldwin
North Vancouver: Gabby Pawlowski
Northampton, MA: Ebru Kardan
Northampton, MA: James Olchowski
Northampton, MA: Megan Labonte
Northampton, MA: Mo Lotman
Nurnberg: Andreas Tobola
Oberlin: Cal Frye
Okinawa: Andrew Scott
Okinawa: Crystal Clitheroe
Okinawa: Maggi Mars Brisbin
Omaha: Courtni Kopietz
Orange County, CA: Emily Sanchez
Orange County, CA: Velvet Park
Orlando: Ida Eskamani
Orlando: Josh Manning
Orlando: Ricardo Williams
Orlando: Valerie Cepero
Oslo: Kriszti Toth
Oslo: Nils Rynning Mork
Oslo: Rita Jonyer
Ottawa, KS: Shawn Dickinson
Oviendo: Miriam Perandones Lozano
Palmer: Jodie Anderson
Palmer: Kalea Hogate
Paris: Emilie Walsh
Paris: Sophia Pagan
Philadelphia: Chris Cummins
Philadelphia: Gina Lavery
Philadelphia: Jill Sybesma

Philadelphia: Michelle Bland
Philadelphia: Simon Joseph
Phoenix: Arianna Guzman
Phoenix: Dan Stone
Phoenix: Serene Dominic
Phoenix: Ty Fishkind
Pittsburgh: Angelos Tzelepis
Pittsburgh: Caitlyn Renee Hunter
Pittsburgh: Francisco Souki
Pittsburgh: Meghan Simek
Pittsburgh: Ralph Crewe
Pittsburgh: Rorry Brenner
Pittsburgh: Sarah Strano
Portland ME: Jen Doebler
Portland, OR: Amanda Thomas
Portland, OR: Chris Trone
Portland, OR: Michael Rainey
Portland, OR: Scott Frey
Porto: Margarida Maia
Pretoria: Janice Laurente
Providence: Russ Beauchemin
Providence: Torrey Truszkowski
Raleigh: Annelise Malkus
Raleigh: Darby Miller
Raleigh: Eric Self
Raleigh: Jeremy Burnison
Raleigh: Johana Bravo de los Rios
Reno: Anna Tatarko
Reno: Louis Grunning
Reno: Valentina Alaasam
Research Triangle Park: Darby Miller
Research Triangle Park: Kerry Donny-Clark
Research Triangle Park: Melanie Clements
Reykjavik: Darius Rodak
Reykjavik: Saori Fukasawa
Reykjavik: Yeonji Ghim
Richmond: Charlie Shanahan
Richmond: Micah Voraritskul
Rochester: Amy Powers
Rochester: Paul Powers
Ruston: Christin Lindley
Sacramento: Kelly Fleming
Sacramento: Logan Hesse
Sacramento: Sarah Light
Saint Cloud: John Mielke
Saint Cloud: Rebecca Woods
Saint Louis: Gary Liming
Saint Petersburg, FL: Brandi Askin
Saint Petersburg, FL: Gerni Oster
Saint Petersburg, FL: Philip Belcastro
San Diego: Brian Rubinstein
San Diego: Lauren Chen
San Diego: Liam Kavanagh
San Diego: Robert Timothy
San Diego: Zed Sevcikova Sehyr
San Francisco: Bart Bernhardt
San Francisco: David Faulkner
San Francisco: Kishore Hari
San Francisco: Lucy Laird
San Francisco: Marciela Abarca
Santiago: Magdalena Cousiño
Santiago de Compostela: Alexandra von Kameke
Santiago de Compostela: Jorge Mira Pérez
Santiago de Compostela: Vanessa Miramontes
Savannah: Angela Coleman
Seattle: Julia Hughes
Seattle: Krunal Desai
Seattle: Marielle McLaughlin
Seattle: Ross Maddox
Sheffield: Kat Hardy
Sheffield: Laura Hayes
Siem Reap: Jennifer Enrique
Silicon Valley: Alex Dainis
Silicon Valley: Andrew Scheuermann
Silicon Valley: Angela Hayward
Silicon Valley: Dan Dobrzensky
Silicon Valley: Danny Haeg
Silicon Valley: Jennifer Fierman
Silicon Valley: Mary Poffenroth
Singapore: Anis Fanaeian
Singapore: Ericka Ward
Spearfish: Garth Spellman
Spokane: Geneva Noel
Stark County: Caiti Waks
Stark County: Megan Pellegrino
State College: Mel Meder
State College: Stephanie Madden
Stockholm: Cheryl Allebrand
Sydney: Amy Reichelt
Sydney: Jessica Grisham
Sydney: Justine Rogers
Sydney: Miriam Capper
Syracuse: Cynthia Handle
Syracuse: Marshall Watt
Tacoma: Laura Moscatello
Taipei: Gigi (Chi-Ching) Chuang
Thunder Bay: Emily Kerton
Thunder Bay: Kaitlin Richard
Thunder Bay: Veronika Swartz
Tokyo: Amanda Alvarez
Tokyo: Andrew Woolner
Tonganoxie: Jake Dale
Tonganoxie: Nicole Holifield
Topeka: Jared Starkey
Toronto: Daniel Moneta
Toronto: Erez Bowers
Toronto: Lauren Shorser
Toronto: Samantha Peck
Toronto: Sarah MacGregor
Toronto: Vanessa Smith
Toronto: Virve Aljas
Tri-Cities, WA: Kat (Kyung-Ah) Park
Twin Cities: David Henry Montgomery
Twin Cities: Jeremy Nilson
Twin Cities: Lindsay Shimizu
Twin Cities: Luke Sharman
Twin Cities: Nicole Bailey
Vancouver, BC: Kaylee Byers
Vancouver, BC: Mike Unger
Vancouver, WA: Amanda Thomas
Vancouver, WA: Leilah Thiel
Vermillion, SD: Jake Kerby
Vermillion, SD: Lindsey Peterson
Vermillion, SD: Mandy Hagseth
Vero Beach: Stephanie Hocke
Vero Beach: Wendy Shafranski
Victoria, BC: Nienke van der Marel
Vienna: Bella Deutsch
Vienna: Christoph Campregher

Wagga Wagga, Australia: Peter Casey
Wakefield: Morgan Haldeman
Waldkraiburg: Dominik Schlund
Waldkraiburg: Herbert Mangesius
Waldkraiburg: Sebastian Eisner
Washington, DC: Aaron Huertas
Washington, DC: Allie Alvis
Washington, DC: Ben Taylor
Washington, DC: Cat Aboudara
Washington, DC: Jenna Jadin
Washington, DC: Rachel Pendergrass
Washington, DC: Tim Starks
Washington, DC: Zaki Ghul
Wellington: Aimee Whitcroft
Wellington: Dell Mitchell
Wellington: Gold
Wellington: Jessica Taylor
Windsor: Ashley Girty
Windsor: Laura Southcott
Windsor: Reg Robson
Winona, MN: Carl Ferkinhoff
Winona, MN: Emily Ruff
Winona, MN: Juandrea Bates
Winston-Salem: Alan Gerlich
Worcester: Ed O'Donnell
Zurich: David Muller
Zurich: Francesca Giardina
Zurich: Jan Schlink
Zurich: Marcella Corti

* It's worth acknowledging that Lewis Weil became a boss of Nerd Nite Austin circa 2009 as a direct result of his drinking too much and subsequently falling off the stage during his presentation about sharks. A natural fit.

Acknowledgments

Adam Durant Birch
Alexander Bailey / DJ Alpha Bravo
Angela Gibson
Anthony Foglia
Ben Lillie (and Caveat NYC)
Ben Wiehe
Brendan T. Kehoe, Esq.
Chris Adams
Claire Cheek
Connie Friess
Dan Rumney
Danielle Svetcov
Dave Balerna (and The Midway Cafe)
Early Nerds: Justin Touchon, Ryan Harrigan, Vikki Rogers, Mike Rush, Rob Worful, Anna Rue, Rui Sasaki, Tim Learmonth, Polly Campbell, and anyone else Chris can't remember because it was too damn long ago
Erin Barker
Francisco Ferrer
Frank Koutis
Gabe Henry, Hope Morawa, Julie Kim, Scott Koshnoodi (and Littlefield)
Irene, John, Kyle, and Lucas Horton

James and Isobel Wasowski
Janice and Robert Girard
Jay Grossen
Jeff Li
Jenn Deguzman
Jessica Girard and Jane
John Karle
Jonathan Bennett
Josh Wolfe
Kishore Hari
Kris Anton
Laura Clark
Laura Jorstad
Layla Yuro
Lee Berman
Mark Rosin
Mary Pilon
Michael Caruso
Michelle Cashman
Mike Kelly
Nettie Wasowski and Matt Marsden
Norma Hoffman
Overlord
Peter Linett
Peter Tatara
Peter Wolverton
Rob Grom
Robert Elmes
Sarah Brockett
Seth Porges
Shyaporn Theerakulstit
The Skint
Sushila and Kumaranayagam Balakrishnan
Vanessa Maruskin